Hands-On Guide to Oscilloscopes

Barry Ross

Hands-On Guide to Oscilloscopes

McGRAW-HILL BOOK COMPANY

London · New York · St Louis · San Francisco · Auckland
Bogotá · Caracas · Hamburg · Lisbon · Madrid · Mexico
Milan · Montreal · New Delhi · Panama · Paris · San Juan
São Paulo · Singapore · Sydney · Tokyo · Toronto

Published by
McGRAW-HILL Book Company Europe
Shoppenhangers Road, Maidenhead, Berkshire, SL6 2QL, England
Tel 0628 23432; Fax 0628 770224

British Library Cataloging in Publication Data

Ross, Barry
   Hands-On Guide to Oscilloscopes
   I. Title
   621.381548

   ISBN 0–07–707818–7

Library of Congress Cataloging-in-Publication Data

Ross, Barry
   Hands-on guide to oscilloscopes / Barry Ross.
   p. cm.
   Includes bibliographical references and index.
   ISBN 0–07–707818–7
   1. Cathode ray oscilloscope.   I. Title.
TK7878.7.R665   1994
621.3815′483– –dc20   94–13387
                                          CIP

1234CUP9654

Typeset by Alden Multimedia
and printed in Great Britain at the University Press, Cambridge.

Contents

Preface

This is a book for anyone who wants to learn about oscilloscopes. All topics of the subject are covered without the assumption of previous knowledge or experience. It deals with how to work an oscilloscope and how an oscilloscope works.

Although the chapters follow a sequence, progressing through the oscilloscope much like a signal does, each chapter is treated as independently as possible. So it should be easy to read any chapter separately, particularly for quick reference.

If you already own or have access to an oscilloscope and some form of signal generator, you will get the best use of this book if you follow through on your instrument. If you do not have a scope, there is a chapter to help you to choose one.

Because of its ability to make voltages 'visible', the oscilloscope is a powerful diagnostic and development tool. The introduction of electronic control systems into all forms of science and industry has meant that the oscilloscope has been introduced to many new disciplines. People involved in medicine, printing, metal fabrication and many other unlikely fields now find themselves needing to operate an oscilloscope in order to control or maintain their own equipment, or to study phenomena which can be displayed through an appropriate transducer as voltage against time.

A spin-off of the rapidly advancing computer industry is the availability of faster and faster integrated circuits. The utilization of these circuits in high speed digital circuitry has generated a series of oscilloscopes with digital storage performances rapidly catching up with those of analog. Undoubtedly in the future, the standard low cost, general-purpose oscilloscope will be an analog and digital storage instrument combining all the advantages of both systems. It is intended that this book will provide the inexperienced user with enough information to understand and operate such a unit.

Many explanations are repeated in different interrelated topics. The reasons for this are twofold:

1 It is intended that different topics can be read in isolation without first having to read all or some of the rest of the book, so many subjects are repeated in different ways as introduction to the topic in question.

2   Several slightly different explanations of the same subject often enable
    people to understand a subject clearly. Where one explanation may not
    seem clear at first, it is hoped that another may enable the reader to cross
    the vital threshold of understanding.

Where front panel controls are referred to, the abbreviated form of the word,
commonly printed on front panels, is highlighted in capitals. For instance
'vertical amplifier VARiable control'. This should help indicate the relationship between these functions in the explanation, and the front panel controls
that select them.

To avoid confusion with respect to the graticule on the CRT screen, note
that when making vertical measurements, the horizontal graticule lines that
are used for reference are those ruled left to right on the screen. When making
horizontal measurements, the vertical graticule lines are used, that is, the lines
ruled vertically on the screen. So vertical amplitude measurements are made
against the horizontal lines across the screen, and horizontal time measurements are made against the vertical lines up and down the screen.

There are application examples given at the end of each chapter. Through-
out the book you can find examples of both how an oscilloscope works and
some practical uses for the instrument. The explanatory figures are contained
within the text, whereas the applications are at the end of the chapters.

The waveform measurements shown in the examples are not necessarily the
settings from the oscilloscope front panel, but include the probe factor. For
instance, if a waveform is measured at 5 V/div, the attenuator may be set to
0.5 V/div, and a ×10 divider probe used. So the actual sensitivity 'at the
probe tip' would be 5 V/div.

List of figures

Glossary of abbreviations

a.c.	alternating current
alt	alternate
a-to-d	analog-to-digital
b/w	bandwidth
Cal	Calibrated
CH1	Channel 1
CH2	Channel 2
CHOP	chopped
CRT	cathode-ray tube
d.c.	direct current
div	division, major segment on graticule
DSO	digital storage oscilloscope
d-to-a	digital-to-analog
EHT	extra high tension (very high voltage)
F	farad, unit of capacitance
FET	field effect transistor
f.m.	frequency modulation
GHz	gigahertz (thousand million hertz), 10^9
GND	ground, earth
ht	high tension (high voltage)
hf	high frequency
Hz	hertz, cycles per second
I	symbol for current
IC	integrated circuit
Intens	intensity
k	kilo ($\times 1000$), 10^3
LCD	liquid crystal display
LED	light emitting diode

LF	low frequency
M	mega (1 million), 10^6
m	milli (thousandth), 10^{-3}
mV	millivolt
n	nano (thousand millionth), 10^{-9}
ns	nanosecond
Ω	ohm, unit of resistance
op-amp	operational amplifier
p	pico (million millionth), 10^{-12}
PC	personal computer
PCB	printed circuit board
PDA	post-deflection anode
p.p.	peak to peak
pF	picofarad
pk	peak
RAM	random access memory
r.m.s.	root mean square
s	seconds, unit of time
s.m.p.s.	switched mode power supply
μ	micro (millionth), 10^{-6}
μF	microfarad
μs	microsecond
V	volts, unit of voltage
X	horizontal axis
Y	vertical axis

1
Introduction

What does an oscilloscope do? It displays a waveform (the picture on the screen) which represents the change in a voltage (usually) with time. There are two kinds of voltage: d.c. (constant) and a.c. (varying). An example of a d.c. voltage is that obtained from an ordinary 1.5 volt battery, and one end is always positive with respect to the other. If the battery voltage were to be measured on a meter, the measurement would be steady (Fig. 1.1). An a.c. voltage does not have a steady value but is constantly changing. If you try to measure a slow moving a.c. voltage on a meter the needle or display will keep moving.

If measurements were taken of an a.c. waveform on a meter once every second, the results might be as follows:

Voltage (volts)	0	0.38	0.70	0.92	1	0.92	0.70	0.38	0
Time (seconds)	0	1	2	3	4	5	6	7	8

Draw a graph of these measurements and the result is half of a sinewave cycle (Fig. 1.2).

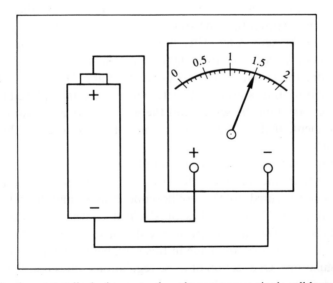

Figure 1.1. Analog meter displaying a steady voltage across a single cell battery.

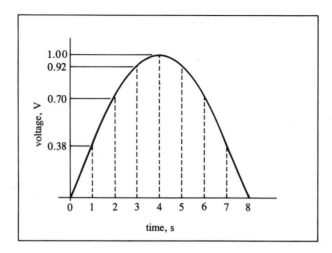

Figure 1.2. Graph showing the amplitude of an a.c. voltage waveform against the base of time.

   The best way to see what is happening to the voltage is not with a meter but on an oscilloscope screen. The oscilloscope creates the time movement across the screen left to right, while the input voltage causes the spot to move up and down and trace out the waveshape automatically. A good way to understand how the oscilloscope works is to set the timebase switch to the slowest speed range and apply a very slow moving input signal. Just intermittently connecting your finger to the input socket will do. Then increase the timebase speed one position at a time and observe the effect on the trace, still applying the same input signal. Eventually the spot moving across the screen becomes a solid line, and continues to be so as the speed is increased further. However, it should now be realized that the spot continues to move rapidly left to right (and back) in the same way. This should help convey the concept that the display shows events from left to right. A waveform starts on the left and as time goes on, the signal is traced out to the right. Hence on a high speed 'solid' waveform, time starts on the left. The voltage is displayed in the vertical direction (axis) on the screen and therefore shows as upward and downward movement. If the voltage goes positive relative to earth the waveform goes upwards.
   The input sockets of the vertical amplifier have an inner 'active' signal input connection, and an outer reference connection. Now this outer connection is usually secured to the oscilloscope chassis and thus becomes the ground connection. When probes or shielded cables are connected to the input sockets, these shields or screens are thus also connected to ground. The majority of oscilloscopes are made in this way, being manufactured to Safety Class I regulations. One section of these regulations states that the chassis and all exposed metal parts must be specially connected to mains safety earth.

There is another class of oscilloscope manufactured to Safety Class II. In this case the reference (ground) point is isolated from safety earth, and these instruments are often supplied with plastic cabinets, and no mains earth connection is made to the instrument at all. Both these classes of instrument are discussed later in Chapter 9. For the moment we shall consider the Safety Class I system, with the chassis and input reference terminal connected to mains earth.

So when using an oscilloscope, displays or measurements are made relative to ground. And if a battery was connected to the Y (vertical) input of the oscilloscope, with the negative side of the battery connected to the ground lead of the probe, the trace or spot would move upwards on the screen – see Fig. 1.3. (The input a.c./gnd switch must be set to d.c.)

The automatic horizontal (left to right) movement of the waveform is caused by the internal circuitry known as the timebase generator. This timebase (sweep) generator causes the spot on the screen to travel from left to right at a precise calibrated speed, and then to return, right to left at a faster speed. (This right to left period is not visible.)

At fast timebase speeds the repeated movement of the spot from left to right makes it look like a line and is known as a trace. The return right to left (flyback) of the trace is not visible as the brightness is reduced or blanked out during this period by the internal oscilloscope circuitry.

The screen of the oscilloscope is marked with a grid known as the graticule, which is usually divided into 1 cm squares. The vertical lines enable time

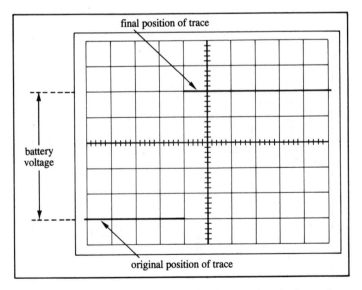

Figure 1.3. CRT screen display showing the deflection produced when a battery voltage is connected to the Y input.

measurements to be made in the X axis, and the horizontal lines enable voltage measurements to be made in the Y axis.

Oscilloscopes are equipped with an intensity control to adjust the brightness of the display, and a focus control to give the best sharpness and definition to the trace. Nearly all the other controls, apart from special functions (such as component tester) are concerned with adjustments to the X and Y display. The basic layout of an oscilloscope is shown in the block diagram in Fig. 1.4.

The power supply system is common to all parts of the oscilloscope, providing the various d.c. voltage supplies required by each section.

The input signal to be examined is connected to the vertical amplifier input socket, and the signal passes through an attenuator switch which can be used to reduce the amplitude of the incoming signal. Attenuate means to reduce, so the attenuator switch can be considered as a 'reducer' switch. From the input amplifier, the signal passes to the vertical output amplifier, and to the trigger circuit.

The vertical output amplifier increases the amplitude of the signal to the relatively large voltages required to deflect the Y plates of the cathode-ray tube.

The trigger circuit converts the vertical input signal from any size or shape of signal to a fixed fast-edged pulse suitable to reliably start the timebase sweep cycle.

The timebase, or sweep generator circuit, provides a ramp or sawtooth-shaped voltage to cause the horizontal deflection plates to scan the cathode-ray tube. Its speed is independent of the input signal, and it can be started or

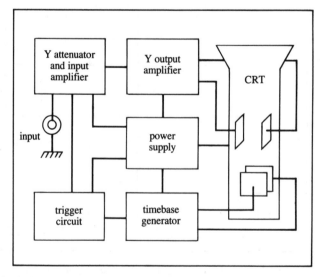

Figure 1.4. Block diagram of basic oscilloscope construction.

triggered by the input signal; and with no input signal connected, it can run by itself.

The X output amplifier magnifies the sawtooth signal from the sweep generator to the large voltages required to drive the horizontal deflection plates of the cathode-ray tube.

The cathode-ray tube itself is the display system for the oscilloscope, generating a narrow beam of electrons and accelerating it forward to provide a spot of light on the phosphor-coated screen. As the beam travels from the cathode at the rear to the front screen, it passes through the control grid for intensity adjustment, and between the two pairs of electrostatic deflection plates – firstly, between the vertical plates and then between the horizontal plates, before striking the phosphor-coated screen, causing it to glow.

We shall look at all sections of the oscilloscope in turn, and see how they work and how they relate to each other.

Figure 1.5 shows the front panel of a typical modern dual trace oscilloscope. Each control is identified, and the key to the figure shows the function and initial set-up conditions. These can be selected by the beginner, and also by the more experienced user when there appears to be a problem.

APPLICATIONS EXAMPLE

Displaying a signal

The object of this example is to display and align a sinewave on the screen. You will need a low frequency sinewave generator to supply a signal of about 6 volts peak to peak, at a frequency of 250 Hz.

Firstly, set the oscilloscope to the automatic trigger condition. There should be a switch in the trigger control area marked 'auto' or 'automatic trigger' which enables you to select manual or automatic trigger. AUTOMATIC trigger should be selected. Set the trigger coupling selector, usually marked a.c., d.c., LF, HF, etc., to the a.c. mode, and the polarity switch to positive (+). Any other trigger functions should be set to normal or off.

The vertical display is usually controlled by a switch marked CH1 only, CH2 only, ALTernate, CHOP, ADD, etc. Set this switch to the CH1 only position. On some models there may be a choice of trigger source from say CH1, CH2, External, Line, etc., so in this case select CH1. On other models, the option may be Internal or External trigger only, so in this case, the trigger source may be automatically chosen by the vertical mode switch as mentioned above. Either way, select Channel 1 trigger.

Now set the vertical and horizontal position controls and the

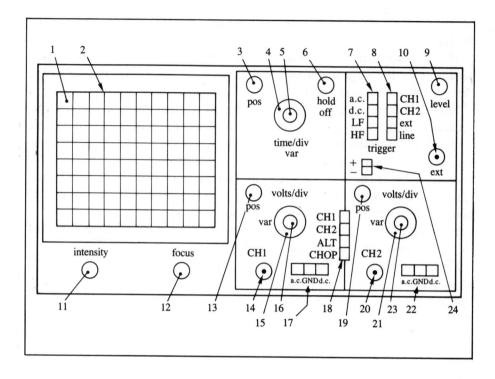

KEY

1 CRT screen
2 Graticule
3 Horizontal position control
   (set to MID-TRAVEL)
4 Timebase switch
   (set to 1 ms/div)
5 Timebase VARiable
   (set to CALibrated position)
6 Variable hold-off
   (set to CAL, OFF or X1)
7 Trigger coupling switch
   (set to a.c.)
8 Trigger source switch
   (set to CH1)
9 Trigger level control
   (set to MID TRAVEL or AUTO)
10 External trigger input socket
11 Intensity control
   (set 75 per cent clockwise)
12 Focus
   (adjust for sharp trace)

13 CH1 position control
   (set to MID-TRAVEL)
14 CH1 input socket
15 CH1 attenuator switch
   (set fully counterclockwise)
16 CH1 VARiable
   (set to CALibrated position)
17 CH1 input coupling switch
   (set to a.c.)
18 Vertical mode selector switch
   (set to CH1)
19 CH2 position control
   (set to MID-TRAVEL)
20 CH2 input socket
21 CH2 attenuator switch
   (set fully counterclockwise)
22 CH2 input coupling switch
   (set to a.c.)
23 CH2 VARiable control
   (set to CAL position)
24 Trigger slope (polarity)

Figure 1.5. Typical front panel layout of an oscilloscope.

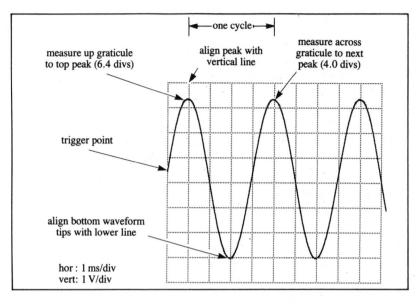

Figure 1.6. Sinewave signal aligned with graticule for amplitude and period measurement.

trigger level control to mid-range. Set the CH1 VARiable (if fitted) and timebase VARiable controls to their CALibrated positions. Set the CH1 attenuator switch to 1 V/div. Set the timebase switch to 1 ms/div. Set the CH1 vertical input coupling switch to a.c. Now you are ready to connect the signal to the Channel 1 input.

Using a screened cable, connect the output signal from the signal generator to the Channel 1 input socket. Adjust the generator output level to about 6 volts p.p. and the frequency to 250 Hz. You should now have just over two cycles of the sinewave signal displayed on the screen, similar to Fig. 1.6.

Now in order to make some waveform measurements, you must position the signal on the screen. Adjust the CH1 vertical position control and align the bottom tips of the waveform with a lower horizontal graticule line, keeping the whole of the signal displayed on the screen. Now measure up the graticule from this horizontal line to the top peaks of the signal. If necessary adjust the horizontal position control so that one of the top peaks is centred on a vertical graticule line. This makes it easier to measure to top peak level as it is then the point where it coincides with the vertical graticule line. In this example, the height from the lower reference line to the top limit of the waveform is 6.4 divisions.

Now adjust the horizontal position control so that the first positive peak is centred on a vertical graticule line. Measure across the graticule from the first peak to the second positive peak. It will be easier if you use the vertical position control to move the display up or down slightly, until the second positive peak touches a horizontal graticule line. Usually only the (vertical) centre line has minor graduations marked on it, so you can use the vertical position control to align the top peaks of the waveform with this centre line. Then you use the graduations on that horizontal line to accurately measure off the distance between the first and second peaks. In the example, this distance is 4.0 divisions.

Now from the results obtained, you can work out the details about the input signal:

The amplitude is 6.4 divisions and the oscilloscope range is 1 V/div, so the signal voltage is 6.4 × 1 = 6.4 volts.

The distance between the first and second peaks of the waveform is 4.0 divisions. The timebase range was 1 ms/div so the period of the waveform is 4.0 × 1 = 4.0 ms. The frequency (f) is the reciprocal (1 over) the period (t).

So

$$f = \frac{1}{t} = \frac{1}{4 \times 10^{-3}} = 250 \text{ Hz}$$

The input signal is thus 6.4 V p.p., 250 Hz.

2
Vertical amplifiers

In the case of a dual trace oscilloscope, we shall assume only one channel is in use, so select channel 1 (or A) only. A voltage signal is connected to the Y input socket. Let us assume that this is a sinewave signal which is varying at 50 cycles per second. One cycle of a waveform is the distance from a given point (A) (see Fig. 2.1) through the variation of the waveform, to that same point when the waveform starts again (B). All oscilloscope voltage measurements are referred to as peak to peak (p.p.). This is the difference between the most positive and most negative points. Now assume that our 50 Hz (50 cycles per second) sinewave input voltage has a size or amplitude of 5 volts peak to peak (5 V p.p.) – see Fig. 2.1.

For the moment, set the input coupling switch (a.c./d.c./GND) to the a.c. position; and the vertical position control to mid-range. Now the amplitude or attenuator switch must be set to a suitable position to give a sensible display on the screen.

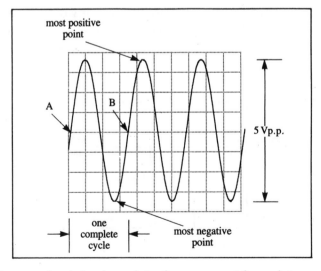

Figure 2.1. Sinewave signal showing points of measurement for peak-to-peak and period measurement.

2.1   Attenuator switch

The attenuator is a multi-position switch which introduces divider resistors
into the path of the input signal to reduce the amount of signal passed on into
the vertical input amplifier. The most clockwise position of the switch is the
'straight through' position where there are no attenuating resistors switched
into circuit. So on this position the same signal that enters the input socket
goes through directly to the input amplifier for that channel. This is the most
sensitive position, typically 5 mV/div.

When the attenuator is set to the most sensitive position, it is effectively out
of circuit. The vertical amplifier usually has a fixed gain or amplification to
make the small signal at the Y input socket large enough to operate the
deflection plates of the cathode-ray tube. So the smallest signal that you can
feed into the oscilloscope input and see displayed clearly on the screen is
determined by this factor. However, if you happen to have a bigger signal
which is too large for the screen, the amplifier gain is not usually altered to fit
it in. Instead the size of the signal is reduced at the input point by using a
potential divider network of resistors. These resistors, with different divider
ratios for each range, form the attenuator switch.

Now the input impedance of the modern oscilloscope is typically 1 Mohm
and 30 pF. What does this mean? Look at Fig. 2.2.

'Looking into' the input socket, what does the signal 'see'? It sees the
internal circuitry of the oscilloscope which, therefore, acts as a load on the
circuit supplying the signal. If you had a signal generator with nothing
connected to its output socket, it would be unloaded. There would be no
signal current flowing out of the generator. But if the socket was connected
via a cable to the input socket of the oscilloscope, it would be loaded by the
1 Mohm resistance and 30 pF capacitance in the oscilloscope. Signal current

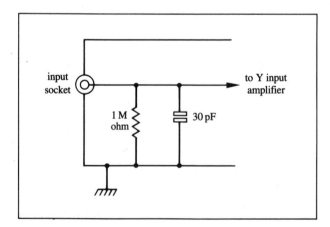

Figure 2.2. Equivalent input circuit of oscilloscope Y amplifier.

would now flow out of the generator, through the cable and through the 1 Mohm resistor and 30 pF capacitance. Now this load is 1 Mohm (i.e. 1 million ohms) with a capacitance of 30 pF (30 picofarads) in parallel with it. The 1 Mohm is a deliberately fitted combination of discrete resistors, but the 30 pF capacitance is largely formed by unwanted but unavoidable stray capacitances. These capacitances are formed between conducting surfaces with small insulating gaps between them, for instance inside transistors, across switch contacts and between copper tracks on a printed circuit board. A small amount of the total 30 pF is sometimes fitted in the form of a variable capacitance trimmer. This is used to keep the input capacitances equal on different channels of the oscilloscope.

The 1 Mohm input resistance is constant for all frequencies. So whether the input signal frequency is 10 Hz or 10 MHz, the loading effect of the resistance is the same. However, with inductances, and more particularly here, with capacitances, the loading effect is directly affected by the signal frequency. The a.c. 'equivalent resistance' for a capacitor is called its reactance, with the symbol Xc. It is calculated from its capacitance value (c), a constant (π) and the frequency (f) at which the reactance is to be determined. The formula for capacitive reactance is:

$$Xc = \frac{1}{2\pi fc}$$

and is measured in ohms, where

$$\pi = 3.142$$
$$f = \text{frequency in Hz}$$
$$c = \text{capacitance in farads.}$$

So, for example, a capacitor of 30 pF (30×10^{-12}) at a frequency of 1 kHz (1×10^3) would have a reactance of

$$\frac{1}{2 \times 3.142 \times 10^3 \times 30 \times 10^{-12}} = 5.3 \text{ Mohm}$$

Now when there is a combination of capacitance (or inductance) and resistance, the effective 'equivalent resistance' of this combination is known as impedance. The symbol for impedance is Z. This combination of resistive and reactive components is what we have at the input socket of the oscilloscope, and on each range of the attenuator switch. Where there is capacitance present across a resistor, the overall impedance of that combination will thus vary with frequency.

In the case of the 1 Mohm and 30 pF input to the oscilloscope, the reactance of the 30 pF capacitor will get smaller and smaller as the frequency increases, and thus reduce the impedance. To overcome this problem, compensating capacitors are fitted where necessary so that the accuracy of the instrument is maintained at all frequencies within its specified range.

Returning to the attenuator switch, let us now turn the knob one position anticlockwise to the 10 mV range (i.e. 10 millivolt per division). This means for every 1 div (or cm) of vertical display height there is 10 mV of voltage at the input socket.

So on this 10 mV position, the signal passed on to the Y input amplifier must be halved (compared to the 5 mV position). However, it is most important that the input impedance of the oscilloscope remains constant (at 1 Mohm/30 pF). So the divider network might look like the circuit in Fig. 2.3.

It can be seen that the total resistance is the sum of the two resistors, 500 kohms + 500 kohms = 1 Mohm as required. In order that the input impedance, and not just the resistance, remains constant, the capacitance of the divider must also be carefully adjusted. Remember that looking into the Y input amplifier there is the total of the stray capacitances mentioned above, and this capacitance is effectively in parallel with the lower 500 kohm resistor shown in Fig. 2.3. It is this capacitance that must be compensated. A variable capacitance trimmer is used to achieve this, and is adjusted during the instrument calibration process to ensure that the attenuator divides by each appropriate ratio over the whole frequency spectrum.

Where the two divider resistors are the same value as here, then the compensation capacitor will be correctly set when its value is equal to the stray capacitance across the lower 500 kohm resistor. Thus if the input impedance is to remain constant at 1 Mohm and 30 pF, the two resistors of 500 kohms plus 500 kohms add to give the total 1 Mohm, and the two capacitors in series must combine to give 30 pF. (In the case of capacitors in series, their values are not added to give the resultant value.) It should be remembered that although the capacitive compensation of this divider ensures that

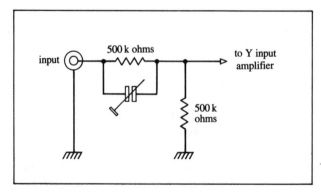

Figure 2.3. Vertical input circuit showing capacitance compensation adjustment.

the 2:1 division is constant for all frequencies, the reactance of the total 30 pF at the input still falls as frequency increases.

Now attentuators switch round in a $1:2:5$ sequence, e.g. 5, 10, 20, 50 mV, etc., and the same principle of divider resistors and compensating capacitors is used on each range to maintain a constant input impedance with the correct dividing ratio.

Now that we have seen how the attenuator switch is used to make the input signal, and hence the display, smaller, let us return to our signal of Fig. 2.1.

2.2   Vertical measurements

To get a suitable display on the screen, we must set the attenuator switch to the most suitable range. In general it is best to have the waveform displayed as large as possible provided it does not exceed the screen height – usually 8 divisions or centimetres. With our input of 5 V p.p., the range 1 V/div would give us 5 divisions of display, i.e.

$$\text{Display waveform } (W) = \frac{\text{Input voltage } (I)}{\text{Sensitivity } (V)}$$

$$\text{So here } W = \frac{5}{1} = 5 \text{ divisions}$$

If we chose the 2 V/div range, we would get $W = 5/2 = 2.5$ divisions, and on 500 mV/div range we get $W = 5/0.5 = 10$ (too large for the screen height of eight divisions).

So the optimum display in this case is the 1 V/div range with 5 divisions peak to peak of vertical display. It should be remembered that the larger the display, the more accurate the measurements that can be made, although in the case of dual trace measurements it is often clearer to see with one waveform at the top and one at the bottom to avoid the traces overlapping.

Attenuators are most commonly 12 position switches, stepping in a 1–2–5 sequence, so with a maximum sensitivity of 5 mV/div, the ranges would be:

$$\underbrace{5 \quad 10 \quad 20 \quad 50 \quad 100 \quad 200 \quad 500}_{mV} : \underbrace{1 \quad 2 \quad 5 \quad 10 \quad 20}_{V}$$

So for a minimum reasonable display of (say) 1 division, a signal as small as 5 mV p.p. can be examined, and at the other extreme, on the 20 V range a voltage as large as 160 V p.p. will give an 8 division display. (Note that these parameters can be extended in both directions by the use of variable controls – gain magnification or attenuating probes – and these features will be described later.)

2.3   Alternating current and direct current coupling

Consider the nature of any voltage waveform. There are three main factors contained in every signal:

1   The size and nature of the voltage variation with time.
2   The speed or rate of the voltage variation with time.
3   The d.c. content of the signal.

Let us consider an example. A simple unregulated power supply might produce a d.c. output voltage of $+100$ V. However, this d.c. voltage may have some 50 Hz mains ripple superimposed on it which we may like to look at and measure on an oscilloscope. We can represent this voltage in two ways:

- On a graph, where zero volts d.c. is shown on the scale. The reference level is the position of the trace on the screen before the signal is connected. So, with the input coupling switch set to d.c. and the input disconnected, the trace line is one division above the bottom of the scale. When the signal is connected, the 100 V d.c. plus superimposed signal, deflect the trace upwards by 5 divisions. This makes the a.c. (signal) comparatively small on the scale of the graph (Fig. 2.4).
- With the d.c. level of the signal removed, so that only the a.c. waveform is observed. Again the reference level is the trace position before the signal is connected. Now with the input switch set to a.c. coupling, the d.c. content is blocked. Now the a.c. signal can be seen in detail, but the display does not keep the d.c. content in perspective (see Fig. 2.5).

Note that the displays in Figs 2.4 and 2.5 are from the same input signal.

In this example, the ratio of d.c. (100 V) to a.c. (1 V) is high, but it is possible to have any combination of a.c. and d.c. in a signal. A voltage could be a high d.c. level with little (as here) or no a.c. signal present; or there could

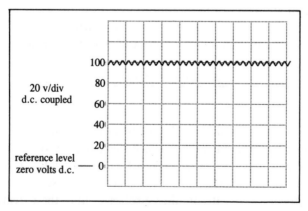

Figure 2.4. Display of an a.c. signal with large d.c. offset; input d.c. coupled.

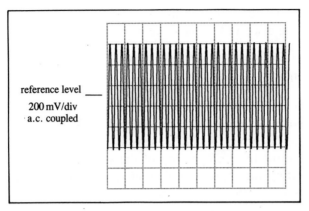

Figure 2.5. Display of the same a.c. signal as in Fig. 2.4, at increased sensitivity and a.c.

be a large a.c. voltage, say 670 volts peak to peak, with no d.c. level present, such as the UK mains supply voltage. These are two 'extreme' cases, but remember that any combination is possible.

Now we shall return to the example in Figs 2.4 and 2.5. Bear in mind that the oscilloscope amplifier has the same magnification for all signals from d.c. to high frequency. As the signal frequency approaches the upper frequency limit of the oscilloscope, the magnification starts to decrease, but for the moment we shall assume that the magnification is constant at all frequencies used. If we are now going to feed in our 100 V d.c. signal and we want a maximum of 8 divisions display waveform:

Since

$$W = I/V$$

Therefore sensitivity

$$V = I/W$$
$$= 100/8$$
$$= 12.5 \text{ v/div}$$

The next higher step on our switch (above 12.5) is 20 V/div, so we select this range and get

$$W = 100/20$$
$$= 5 \text{ divisions}$$

So now when we connect our signal into the scope, if we select d.c. on the input coupling switch, the 100 V d.c. voltage goes into the amplifier and the trace will move up the screen 5 divisions. If our ripple (a.c.) voltage is say 1 V p.p., then the displayed ripple waveform will be

$$W = 1/20$$
$$= 0.05 \text{ divisions}$$

In other words our ripple waveform is only 0.5 mm high (peak to peak display size) on the screen. So in order to examine and measure the ripple voltage, the attenuator must be turned to a higher sensitivity. But to avoid the (100 V) d.c. content moving the trace off the top of the screen, we must 'block' the d.c. content by setting the input coupling switch to a.c. This will introduce a capacitor in series with the input signal, which will allow the a.c. ripple voltage into the amplifier, but block the d.c. voltage from getting through. So now we can select, *say*, the 200 mV/div range and get 5 divisions of displayed ripple waveform.

It should be noted here that there is a limit to the d.c. voltage that can be blocked by a capacitor, usually about 400 V. This limit must include the peak a.c. as well as the d.c. voltage.

If the maximum permitted input voltage is exceeded, the input circuitry can be severely damaged. Vertical input amplifiers are normally fitted with over-voltage protection circuitry, but this can only work up to a certain level. It is always possible to damage the input circuit if the applied voltage is large enough, so no matter how good the protection, it has its limit. The use of a ×10 divider probe is recommended at all times as this will increase the input protection, but with or without a probe, the maximum input voltage rating must be observed.

It is good practice to use the a.c. position and set the attenuator to the least sensitive position (e.g. 20 V/div) before connecting to unknown signals.

So it can be seen that the a.c./d.c. switch can be used to make two alternative measurements of the same signal. If the switch is first set to the GROUND position, then the trace aligned with a horizontal graticule line using the shift or position control; then d.c. coupling selected, the trace movement can be measured as the d.c. content of the signal. If a.c. is now selected and the attenuator sensitivity increased (if necessary), the a.c. 'signal' part of the waveform can be examined and measured.

It should be noted here that there are a few points to bear in mind regarding the a.c. and d.c. coupling:

2.3.1 *On a.c. coupling*

1   There is a low frequency limit to the signals that can be examined on a.c., usually about 2 Hz, so at low frequencies approaching this limit, take care with amplitude measurements (if possible use d.c.). For signals below

20 Hz, amplitude errors greater than 3 per cent will occur, and these errors get larger as the frequency reduces. At about 2 Hz, the error will be 30 per cent at the $-3\,$dB point.

2 When observing rectangular or pulse type signals, the vertical position of the trace will depend on the mean level of the signal. So if the mark to space ratio of the signal changes, the trace will move on the screen. The mark to space ratio is the ratio of the width (time period) of the positive going part of the cycle to the negative going part of the cycle (see Fig. 2.6).

For all of the waveforms shown in Fig. 2.6, the trace is set to vertical screen centre before applying the input signal.

When the signal is more positive than negative as in the top waveform in Fig. 2.6, the mean level of the signal is positive and so the trace moves upwards on the screen. This is where the duration of the positive half-cycle is greater than the duration of the negative half-cycle. With the opposite case, as at the bottom of Fig. 2.6, the duration of the negative half-cycle is greater, and so with a negative mean level, the trace moves downwards on the screen. With an equal mark to space ratio, the positive and negative duration is the same, so the waveform is equally disposed about the reference line.

As well as moving the reference position of the trace, it may also cause loss of triggering. Again use of d.c. coupling will eliminate this effect and allow the vertical trace position to remain constant.

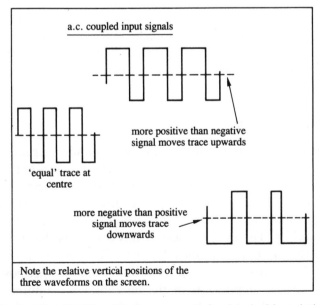

Figure 2.6. Display showing the vertical movement of a signal with variation of mark : space ratio when a.c. coupled.

2.3.2   *On d.c. coupling*

Signals with large d.c. offset may cause the trace to disappear from the screen and make it difficult to find – particularly when manual trigger is in use. Remember that the input voltage may contain any combination of a.c. signal and d.c. level. The d.c. level alone may cause the trace to move right off the top (positive d.c. content) or bottom (negative d.c. content) of the screen. Also the d.c. screen movement may move the trace beyond the trigger level range even though the trace is still within the screen area. Once the signal is outside the trigger level range on manual trigger, the timebase stops, so there is no sweep at all. In this case it is helpful to momentarily switch the input coupling to the a.c. position to establish exactly what the input signal consists of, and then revert to d.c. coupling if required.

2.4   Vertical position control

This function operates by introducing a d.c. offset voltage within the oscillo-scope amplifier and thus moves the vertical position of the trace on the screen. The amount of movement available is usually two to three times the screen height, that is say 20 divisions of shift, so quite large amounts of d.c. content in the input signal can be offset by the position control. The position control does not have any effect on the input signal itself but it does set the reference position of the trace. So when making d.c. measurements on the screen, the position control must not be altered after the initial set-up reference position. Subsequent alteration of the position control could then be mistaken for a d.c. change in the displayed signal. In general it is much better to use a d.c. voltmeter for d.c. measurements where there is no a.c. signal involved. They are more accurate and easier to use for that particular job.

2.5   Variable Y gain

The VARiable gain control gives continuous variation of gain between any two attenuator ranges. Since the attenuator divides the signal in a fixed 1, 2, 5 sequence, the largest change is from the 2 to 5 positions where the attenuation changes by $5/2 = 2.5 : 1$. So the VAR gain control always provides just over a 2.5 gain range, and thus provides continuous overlap of all attenuator ranges on the Y amplifier. The disadvantage of this control is that it has only one calibrated position (CAL). It then has no effect on the attenuator scale reading, in other words it is in the 1 : 1 position. In any other position of this control, the vertical amplifier is in the uncalibrated condition, so voltage measurements cannot be made. So although the variable control is very useful for adjusting the display height to a convenient or exact size (see risetime measurement), you *must* set it to the CALibrated position before making measurements.

2.6   Dual trace

So far we have dealt with the controls and operation of a single vertical amplifier. But the vast majority of oscilloscopes available are dual trace instruments. All the above controls and facilities are duplicated to give two traces on the screen, usually Channel 1 and Channel 2 or A and B. Although it appears that there are two traces displayed simultaneously on the screen, this is an illusion. In the same way as the single spot rapidly moving left to right appears to be a solid line, two inputs either chopped or alternately displayed across the screen appear like two simultaneous lines. The two ways of displaying the traces, in CHOPped or ALTernate mode, are usually selected by the operator, although some oscilloscopes do it automatically with operation of the timebase switch.

2.7   Alternate mode

If the timebase switch is set to a slow speed, say 50 ms/div or slower, it is easy to see the alternate mode in operation. With both vertical inputs set to ground position, set CH1 position to the top half of the screen and CH2 to the bottom half. Now with each sweep of the timebase, alternately CH1 and then CH2 are displayed. The slower the timebase speed, the easier it is to see.

The signals from the CH1 and CH2 pre-amplifiers are fed through a switched 'gate' to the output amplifier. This gate is switched over once each time the sweep ends. So during the flyback period of the timebase, control switches from CH1 to CH2 to CH1 to CH2, etc., once for every sweep. As each channel is gated through to the output amplifier, so its signal and position are displayed on the screen. The changeover is designed to take place in the flyback period because then the spot is blanked and not visible, so the CH1 to CH2 transition is not seen. Once the timebase is switched to high speeds, the two alternate displays cease to be discernible, so they appear as two solid continuous traces.

Although for the majority of applications it is unimportant, it is worth remembering that the two traces are not actually there at the same time. So when different signals are displayed on each channel, they are *not* displayed together in real time. For repetitive signals it does not matter, but for intermittently occurring pulses it may be very important.

It should be clear from the above description that the alternate mode has a big disadvantage. On slow sweep speeds, it ceases to perform the dual trace function. Only one trace is displayed at a time, so another technique is used to obtain two traces at slow speeds. This is known as the CHOP or CHOPped mode.

2.8   Chop mode

The control gate feeding *either* the CH1 *or* CH2 signal to the output amplifier in the alternate mode is again used. As stated above, in alternate mode, the CH1 then CH2 signals are fed through to the CRT at the end of each sweep of the timebase. Each channel is displayed for one complete sweep.

Now in the chop mode an independent, high frequency oscillator is used to switch the gate. That is, instead of the CH1 and CH2 signals being switched through to the output amplifier (and hence the screen) once every sweep cycle, the switchover takes place many times during one sweep cycle. As the spot moves left to right across the screen, it displays CH1, CH2, CH1, CH2, etc., across the screen. The rate at which this switchover occurs is determined by the independent clock oscillator. It usually runs at about 1 MHz. Now at a slow sweep speed of, say, 10 ms/div, the duration of a 10 division sweep is 10×10 ms $= 100$ ms, one-tenth of a second. So the chop oscillator running at about a million times a second (1 MHz) will switch many times during one sweep.

The period of the 1 MHz oscillator will be $1/f = 1$ microsecond or 1×10^{-6} s. So in 100 ms (100×10^{-3} s) there are $100 \times 10^{-3}/1 \times 10^{-6} = 100 \times 10^{-3}$ (100 000) switched operations.

So the gate will switch between CH1 and CH2 100 000 times during the sweep. Thus CH1 and CH2 are displayed 50 000 times each. So at these slow sweep speeds, CH1 and CH2 each appear as continuous traces. The display time of each channel is long compared to the switching transition time between each channel, so the switchover transitions are barely visible. Just to make sure, a blanking signal is usually fed to the CRT to blank the display during the transition period.

So now the oscilloscope can perform the dual trace function at all speeds of the timebase. The disadvantage of the chop system is that at fast timebase speeds, the duration of the sweep becomes shorter than the period of the chop oscillator, so the dual trace function can cease. So both alternate and chop modes are included for most appropriate use according to sweep speed. On some manufacturers' oscilloscopes, the ALTernate/CHOP switchover is done automatically by the timebase switch. Although this simplifies the operation of the instrument, it does remove the choice of when to select each mode, and in some cases it is advantageous to choose a specific mode together with a particular timebase speed. There are oscilloscopes with multitrace systems displaying between three and eight traces using the same chop and alternate principles as above. These instruments are particularly useful for displaying digital waveforms where there may be (say) eight separate pulse waveforms making up an eight bit word.

2.9   Invert

The invert function, as the name suggests, reverses the polarity of the input signal waveform on the screen. It displays the signal upside down. When the invert function is operated, a signal that is positive going with respect to ground reference will go downwards on the screen and vice versa. Instruments may have invert switches fitted to either or both vertical channels, but the main virtue of the invert function is not the polarity reversal in itself. It is rather for use in conjunction with the ADD facility usually also fitted. It then enables the oscilloscope to be used as a differential amplifier, with the CH1 and CH2 inputs as the differential inputs.

2.10   Add

Referring back to the chop and alternate modes, each input channel, say Channel 1 and Channel 2, was sequentially switched through to the vertical output amplifier and CRT plates. So signals entering each input were alternately displayed on the screen, either many times each per sweep, or alternately once each sweep.

Now it is possible for the gate to pass both the Channel 1 and Channel 2 signals through to the CRT simultaneously. This is the ADD or SUM mode. Only one waveform appears on the screen, and this is the combination of both input signals added together. For example, both channel inputs may be connected to the same sinewave source and displayed on the screen in chop mode as two traces each two divisions peak to peak. Now select ADD mode. The combined display now gives four divisions peak to peak.

Select the invert function on one channel and the display should reduce to a line. The inverting of one channel has reversed its polarity. So now one positive signal is added to an equal negative signal, and they cancel each other. Hence the line. Now if there is a difference between the signals in each channel, this difference will appear instead of a line. Where the signals are identical they cancel. Where there is a difference it is displayed. Hence this is known as the differential amplifier mode. By using this differential measuring technique, it is possible to measure accurately very small difference signals between two much larger signals (or d.c. voltages). Figure 2.7 shows two waveforms in the dual trace chop mode. The upper trace shows the signal input to an operational amplifier, and the lower trace shows the output. The upper trace (CH1) sensitivity is set to 0.2 V/div, the lower (CH2) to 2.0 V/div, and the op-amp has a gain of about 10. Since the sensitivity of CH1 (0.2) is 10 times greater than CH2 (2.0) the two waveforms should be identical except for

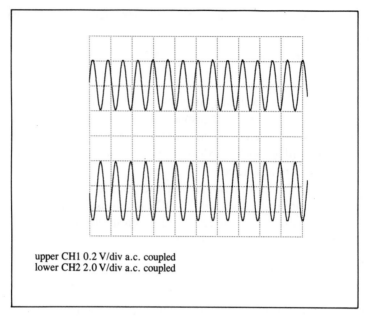

upper CH1 0.2 V/div a.c. coupled
lower CH2 2.0 V/div a.c. coupled

Figure 2.7. Display showing the input and output signals from an operational amplifier.

the phase reversal due to the op-amp. That is, the two waveforms on the screen should have exactly the same amplitude, and if added together should cancel each other out.

Now when ADD is selected on the oscilloscope, the result is the difference between them as seen in Fig. 2.8. It can be seen that in Fig. 2.7, the CH2 output waveform is slightly larger than the CH1 input, so the gain is greater than 10. Hence there is a differential signal displayed in Fig. 2.8. In this case it is not necessary to use the INVERT function, as the two signals are already antiphase due to the inverting action of the op-amp.

2.11   Vertical main amplifier

Having looked at the basic control functions of the vertical amplifier, let us now look further into the system. The signal emerging from the output of the attenuator is fed into a multi-stage amplifier system which finally drives the CRT plates. These amplifier stages are always d.c. coupled and always end up at the output as a push pull drive to the CRT plates. Most wideband oscilloscopes use electrostatically deflected CRTs with a pair of opposing plates for the y deflection. Push pull means that two opposing signals are generated, working in antiphase, so that when one signal voltage is positive going, its identical counterpart is negative going (see Fig. 2.9).

The negative going voltage will repel the electron beam passing through the CRT, while the positive going signal will attract the beam. The 'push' effect

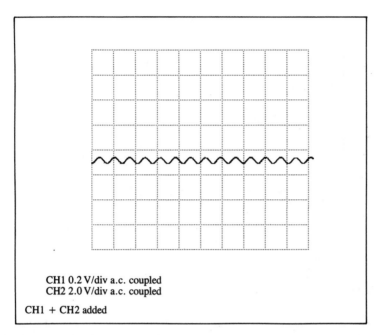

CH1 0.2 V/div a.c. coupled
CH2 2.0 V/div a.c. coupled

CH1 + CH2 added

Figure 2.8. Display showing the use of ADD function to cancel the two antiphase waveforms of Fig. 2.7.

on one plate and 'pull' on the other effectively doubles the amplification of the amplifier (as well as other beneficial effects).

Electrostatic deflection refers to the way that the X and Y plates work. There is a pair of rectangular metal plates, one above and one below the electron beam axis; and another pair, one each side of the beam axis. When a voltage is applied across a pair of metal plates, the electric charge on each plate sets up an electric (or electrostatic) field between them. When the electron beam from the cathode passes through this electric field it is influenced (deflected) by it.

Another, perhaps better known, field system is the electromagnetic field which is caused by a current flowing through a conductor, usually a wire. This is found for instance in electric motors, and television tubes, where a wire coil is fitted around the neck of the tube. In this case the scanning signal is fed to the deflection coil, and the resultant electromagnetic field deflects the beam.

In this book, all references to CRT deflection will assume electrostatic deflection.

There are three parameters associated with the Y amplifier system.

2.12   Sensitivity, accuracy and bandwidth

These parameters refer to the whole of the Y system from the input socket to the CRT display. Thus the input attenuator already discussed is included.

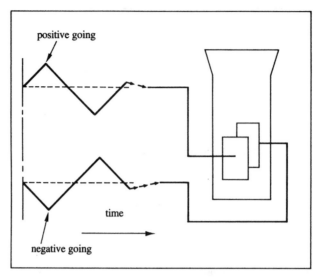

Figure 2.9. Diagram showing two antiphase, push pull signals fed to CRT vertical

2.12.1 Sensitivity

The sensitivity is the amount of voltage required at the input socket to deflect the trace by exactly one division. The figure usually given in specifications is the basic calibrated sensitivity where the full bandwidth is available, and does not include use of magnifiers or variable controls. The sensitivity tells you just how small a voltage signal you can display on your oscilloscope and get one division of display.

2.12.2 Accuracy

The accuracy is the tolerance of the calibrated measuring capability, typically 3 per cent. So if a waveform is measured on the screen as 5 divisions of vertical height (peak to peak) on a range of 1 V/div, we must expect a possible error of 3 per cent of the 5 divisions (5 volts). That is, if the measured amplitude is 5 divisions, the actual signal could be 5 V plus or minus 3 per cent of 5 V.

$$5 \text{V} \pm (3/100 \times 5)$$
$$= 5 \pm 0.15$$
$$= 4.85\text{--}5.15 \text{V}$$

So due to the possibility of a 3 per cent error, a signal measured as exactly 5 V might actually be somewhere in the range 4.85–5.15 V.

Alternatively, if the actual signal is exactly 5 V, the display on the screen could be 4.85–5.15 divisions. In practice the errors should be much less than this since the specification limits are worst case and not typical.

2.12.3 Bandwidth

The bandwidth of the vertical system is the range of frequencies that it can handle. One of the main limiting factors affecting the amplifier is capacitance. Any two conducting surfaces will form a capacitance between them, particularly where the gap is very small. So inside transistors, inside integrated circuits, between copper tracks on a printed circuit board (PCB), etc., will all form capacitances. At low frequencies, these capacitances have little effect on the circuit, but at high frequencies they become noticeable. The a.c. 'resistance' or reactance of capacitors reduces as frequency increases, so the amplifying circuit tends to be loaded down by these reducing reactances at high frequency. The effect is to reduce the gain of the amplifier. Conventionally the bandwidth is given as the minus 3 dB point (−3 dB). This is the frequency where the gain of the amplifier has reduced to 0.707 times the low frequency gain (0.707 = 1/root 2, square root of 2 = 1.414). See Fig. 2.10.

Suppose we connect an input sinewave of 5 V p.p. to the oscilloscope, with the attenuator on the 1 V/div range. Let us suppose that the bandwidth of the oscilloscope is 20 MHz. Now if we set the sinewave generator frequency to 1 kHz, the display will be 5 divisions. If we now increase the sinewave generator frequency to 20 MHz (keeping the voltage supply constant at 5 V p.p.), the display will reduce to 0.707 × 5 = 3.5 divisions (approx.). So, as the input signal frequency approaches the oscilloscope bandwidth, amplitude measurements become unreliable. If the input signal is anything other than a

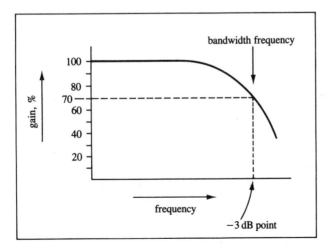

Figure 2.10. Graph showing the gain/frequency response of an oscilloscope vertical amplifier, with −3 dB bandwidth point.

sinewave the situation becomes even worse. What can we learn from all this? If the signal to be observed is sinusoidal, the oscilloscope should have a bandwidth at least twice the input frequency for reliable measurements. Even so, at half bandwidth frequency, the amplitude error is approximately 10 per cent. If the input signal is non-sinusoidal, the bandwidth should be 10 times the input frequency (as a general guide).

If the frequency of your input signal goes anywhere near the bandwidth of your oscilloscope, take extra care, the signal displayed on the screen may not be a true representation of the signal going in the input socket. If the signal is a sinewave, the amplitude may be too small. If the signal is non-sinusoidal, its shape may be distorted with sharp corners being rounded off and narrow pulses being attenuated.

2.13   Delay lines

When a low frequency waveform is displayed on an oscilloscope the left-hand side start of the display is chosen by the trigger level control or the auto trigger circuit. Let us assume that this low frequency signal is a 1 kHz sine-wave. Now increase the input signal frequency to 1 MHz and increase the timebase speed accordingly to get the same displayed waveform as before. It will be seen that less of the first cycle is observed at the start of the waveform. The trigger point appears to have moved up the waveform. If the input waveform is switched to a square wave, it will be seen that the display starts at the top of the leading edge or even on the flat top of the waveform.

These problems are caused by the start-up time of the timebase system, and can be overcome by the introduction of a signal delay line. When the input signal goes into the trigger circuit, it takes a finite time to get the sweep system started, and the spot moving across the screen. During this small time, usually about 50 ns, the spot is still blanked and therefore not visible.

Even though the input waveform has crossed the trigger point threshold, and is continuing its vertical excursion, it cannot be seen until the sweep system has started and the spot is unblanked. Although this time delay is very small, and insignificant at low frequencies, it means missing a large (and possibly vital) part of the display.

Now a delay line is a special cable in the vertical amplifier through which the signal passes on the way to the vertical deflection plates in the CRT. It is connected in the vertical amplifier *after* the point where the signal is routed to the trigger circuit (Fig. 2.11). The signal takes a long time (relatively speaking) to get through the delay line, and reach the Y plates of the CRT (usually about 200 ns).

While the signal is passing through the delay line, it meanwhile triggers the sweep; the X amplifier starts moving the spot; and the unblanking circuit

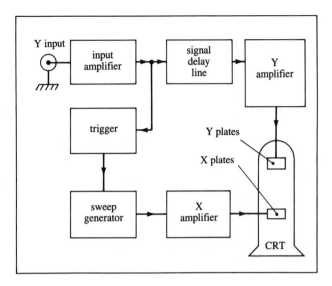

Figure 2.11. Block diagram showing the relative position of the signal delay line in the oscilloscope.

makes the spot visible. So by the time the waveform reaches the Y plates, the trace is already moving and visible before the trigger threshold point is reached on the screen.

In the case of a square wave or pulse, it is now possible to see not only the whole of the leading edge, but even the flat top (or bottom) of the previous half-cycle. (This is provided that the pulse or square wave has a risetime less than the 200 ns duration of the delay line.)

Figure 2.12 shows the time relationship between the signal, sweep and display, with and without a signal delay line. The top diagram shows the input waveform and the second shows the position of the trigger pulse which is at the same point in time as the start of the waveform display. The third diagram shows the sweep waveform relating to the input waveform, the sweep start being coincident with the trigger pulse which initiated it. Due to the time delay for the sweep to start and the spot to become unblanked and hence visible on the screen, the input waveform is not visible until a short time after the trigger pulse occurs. This is shown in the fourth diagram where the front edge is missing. The final diagram shows the displayed picture when a delay line is fitted. The 200 ns delay introduced by the delay line causes the signal to arrive at the Y plates a little later, by which time the timebase has started and the spot is visible. Now when the leading edge of the waveform rises, it can be seen on the screen as shown.

The delay line is in fact a twin conductor cable, each conductor used in circuit to delay the the two antiphase push pull signals. It is wound in a special way to give maximum performance, having to add a time delay to all

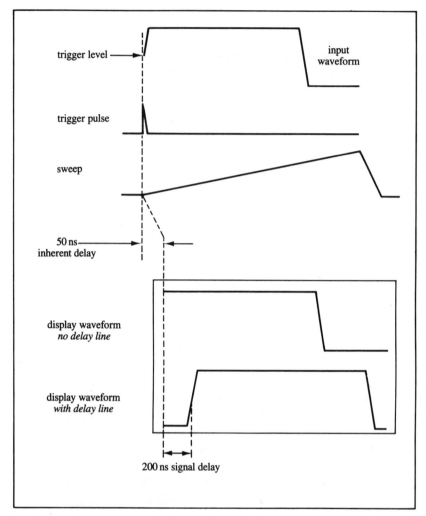

Figure 2.12. Waveforms showing an input signal displayed without and with use of a signal delay line.

frequencies of signal passing through it, while at the same time having a wide bandwidth to allow the high performance of the vertical amplifier. The two push pull signals are fed from the delay line into the vertical output amplifier, to provide the high frequency, high voltage drive for the CRT Y deflection plates.

In digital storage oscilloscopes, a pretrigger system is used to achieve the same effect as the signal delay line. Digital circuit techniques are used to give a display which includes events before the trigger point on the signal: the 'pretrigger' events.

So the analog delay line and the digital pretrigger system both allow the display of signal activity just prior to the trigger point.

APPLICATION EXAMPLE

Measurement of transistor amplifier gain

The object of this example is to measure the gain of an NPN transistor voltage amplifier. A sinewave signal is fed into the amplifier at a frequency of 1 kHz. The signal going into the base of a transistor stage is measured on the upper trace (CH1) of the oscilloscope, and the magnified signal appearing at the collector is measured on the lower CH2 trace. The transistor action causes the collector signal to be inverted compared to its base input signal, so the upper and lower traces are seen in antiphase. Since we are concerned only with the signal amplification and not the d.c. levels, CH1 and CH2 input coupling switches are set to the a.c. position. Consequently the displayed waveforms only show amplitude and shape, not d.c. content. Figure 2.13 shows a part of the transistor amplifier concerned. The transistor stage T2 is an NPN, common emitter amplifier, and we shall measure the stage gain from base to emitter.

Connect the CH1 probe to the base of T2, and the ground reference to earth. The CH2 probe is connected to the collector of T2. Now adjust the output sinewave from the signal generator until there is an

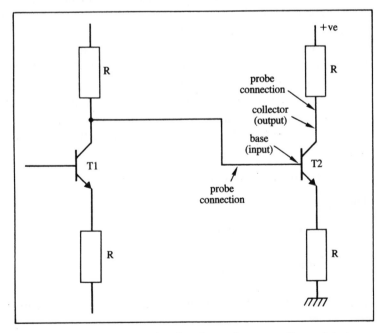

Figure 2.13. Part of circuit diagram showing the probe connection points to measure transistor amplifier gain.

Hands-on guide to oscilloscopes

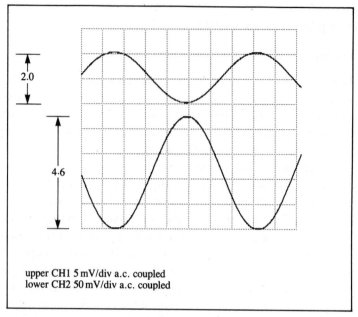

upper CH1 5 mV/div a.c. coupled
lower CH2 50 mV/div a.c. coupled

Figure 2.14. Screen waveforms obtained before and after transistor amplification.

exact number of whole divisions on the screen. In this case, the input at the base of T2 is set to give a 2.0 division display (see Fig. 2.14, CH1). The CH1 VARiable is set to the CAL position, and the sensitivity noted as 5 mV/div, using a ×1 probe.

The CH2 probe is connected to the T2 collector, and Fig. 2.14 lower trace shows the signal output. With the CH2 VARiable set to the CAL position, and a ×1 probe used, the sensitivity is noted as 50 mV/div. Now the peak-to-peak amplitude of the CH2 display is measured against the graticule. The result is 4.6 divisions.

The stage gain for T2 is the output divided by the input, so here the output is:

$$4.6 \text{ (divs)} \times 50 \text{ (mV/div)} = 230 \text{ mV}$$

The input was

$$2.0 \text{ (divs)} \times 5 \text{ (mV/div)} = 10 \text{ mV}$$

$$\text{Stage gain } A = \frac{230}{10} = 23$$

Thus in this circuit, the T2 transistor stage has an amplification of 23.

3
Probes

The problem of actually connecting your signal to the oscilloscope is best overcome by the use of a probe. But which is the best probe to use? When you connect the tip of the probe to some part of a circuit to look at the signal at that point, you immediately change or modify the conditions at that point. This is not a situation to cause alarm! The change to the circuit under test is probably (and usually) very slight. Ideally the probe would not affect the circuit it examines but in practice this is impossible. However, we should aim for this objective by making the probe as unnoticeable to the circuit as possible. This means using a high impedance probe whenever possible. High impedance means high input resistance and low input capacitance at the probe tip. So in general a $\times 10$ probe is recommended as the best compromise. In the case of probes, 'times 10' refers to the impedance of the probe, it *does not* magnify the signal by 10. In fact quite the opposite, the $\times 10$ probe actually divides the signal amplitude by 10 and a $\times 100$ probe divides by 100.

Because probes are usually quite inexpensive items compared to the cost of an oscilloscope, it does not mean that they are unimportant. On the contrary, the probe specification is just as important as the scope specification. Any signal displayed on the oscilloscope screen also passes through the probe first, so if the probe were to modify or distort the signal in any way, the oscilloscope could not display the true signal. So it is worth looking after the probes, and periodically checking their performance and compensation adjustment to ensure that they are in optimum condition.

As stated above, any signal going into the oscilloscope passes through the probe first, so the probe bandwidth will combine with (and reduce) the oscilloscope bandwidth. Therefore, the probe bandwidth must always be higher than the oscilloscope bandwidth. If possible, choose the probe bandwidth to be at least 10 times that of the oscilloscope (see Chapter 8, Sect. 8.21).

On most oscilloscopes you will always have to remember which type of probe you are using, and make the appropriate correction to the sensitivity or actual volts per division at the probe tip. On some higher performance (higher cost) models, however, the probe reduction factor is automatically corrected by the oscilloscope. This automatic probe correction will only work if the scope manufacturer's own recommended probes are used. This is because the probe connector has special contacts which the scope input socket detects, and hence recognizes the probe type. When one of these coded probes is connected to the oscilloscope various factors may be affected. These may include (for each channel): the attenuator range indicator; screen readout display indication; vertical cursor measurement factor and peripheral transfer data, e.g. data bus to a computer, printer, etc.

For the majority of oscilloscopes, you will have to make the mental correction for the probe factor each time you make a voltage measurement.

3.1   ×1 Probes

We saw that the oscilloscope vertical amplifier usually has an input imped-ance of 1 Mohm and 30 pF (approx.). If we now use a ×1 probe to connect the signal to the input socket, the input resistance at the probe tip remains constant at 1 Mohm since there is no attenuating resistor built into the probe. However, the input capacitance will be the added values of the probe cable capacitance and that of the oscilloscope itself. This will usually give a total value at the probe tip of about 50 pF. As the frequency increases, capacitive reactance or 'a.c. resistance' reduces. At the bandwidth frequency of an oscilloscope of 20 MHz for instance, the reactance of this 50 pF capacitance is only about 160 Ω. So at this frequency the probe will only have an input impedance of about 160 Ω compared to 1 Mohm at low frequency. SO BEWARE! This type of probe should only be used for very low frequency work up to (say) 100 kHz, or when the full ×1 sensitivity of the oscilloscope is absolutely needed.

3.2   ×10 Probes

Now consider the use of a ×10 probe. The probe body contains a very high resistance of 9 Mohms, which with the 1 Mohm input resistance of the oscil-loscope forms a total input resistance of 10 Mohms, and a resistive divider to the signal of 10:1. In order that the input impedance of the probe remains constant over the frequency range, a variable compensating capacitor is connected across the 9 Mohm resistor. This must be adjusted by the user to optimum setting (see Fig. 3.1).

Since the reactance of the 30 pF input capacitance of the oscilloscope and the probe cable capacitance get progressively smaller as frequency increases, the effective input resistance is reduced from 1 Mohm to a lower and lower value as the frequency gets higher and higher. So what was a 10 : 1 divider at low frequency, becomes 20 : 1, 40 : 1, etc., as the frequency goes up. To overcome this problem a compensating capacitor across the 9 Mohm resistor progressively reduces the effective resistance of this part of the divider as frequency increases. When this trimmer capacitor is correctly adjusted, the

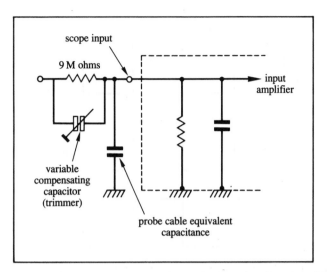

Figure 3.1. Circuit diagram showing ×10 probe equivalent circuit and oscilloscope input circuit.

10 : 1 divider ratio of the probe is constant over the whole input frequency range.

The normal method of adjusting the probe is to connect the probe tip to a square wave signal generator. This should have a frequency of approximately 1 kHz and a fairly fast risetime. The capacitor trimmer in the probe body is then adjusted to give a square corner to the leading edge of the square wave (see Fig. 3.2).

If the capacitance of the trimmer is set too large (a), the attenuation of the probe will be less than 10 : 1 at high frequency. The high frequency 'component' of the waveform is at the front (left) edge so this part of the waveform will receive less attenuation than the rest and result in 'overshoot' (a). If the trimmer is set too small, the square wave will have 'undershoot' (b). Here the input capacitance of the cable and oscilloscope will cause the high frequencies to be attenuated more than the rest. Correctly set, the square wave will be square (c)!

Now the compensating capacitor is effectively in series with the input capacitance of the oscilloscope (plus cable) (Fig. 3.3). This has the beneficial effect of reducing the input capacitance at the probe tip by a large factor. Typical values of input capacitance for ×10 probes are 10–15 pF. So the effective loading that this type of probe has on a circuit is typically 10 Mohm and 12 pF (see Fig. 3.4).

The great disadvantage of the ×10 probe is its attenuation of the signal amplitude. For one thing the basic sensitivity of the oscilloscope is reduced from (say) 5 mV/div to 50 mV/div when the probe is used. In most cases this is not too worrying as most signals are greater than 100 mV or so.

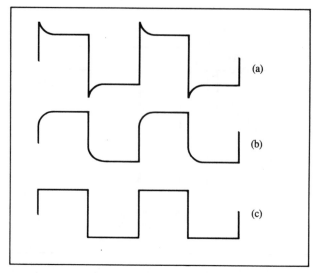

Figure 3.2. Square wave response of a ×10 probe incorrectly and correctly adjusted.

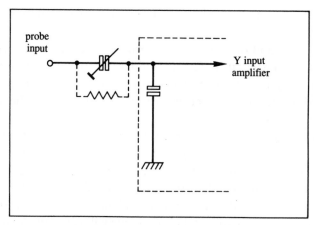

Figure 3.3. Equivalent circuit of probe compensation capacitance in series with oscilloscope input capacitance.

The other problem with this probe is always remembering to make the correction to the vertical sensitivity, no matter which attenuator range you are using. However, if a ×10 probe is always used, dividing the sensitivity by 10 soon becomes a habit.

The input capacitance may vary between one channel and another on an oscilloscope; it may be (say) 29 pF on Channel 1, and 30 pF on Channel 2. This small difference will make an enormous difference to the probe compensation, so if you move a probe from one channel to another, always check the compensation trimmer (see Fig. 3.2). Moving a probe from one

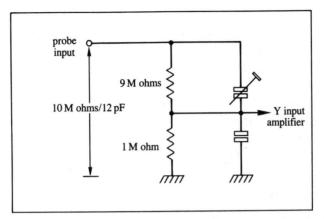

Figure 3.4. Equivalent circuit of probe and oscilloscope input impedances.

oscilloscope to another will require its trimmer to be reset each time. Many oscilloscopes have their two (or more) inputs matched. Channel 2 (say) may have an extra capacitance trimmer across its input to allow the Channel 2 input capacitance to be set exactly the same as Channel 1. Then, of course, moving a probe from one channel to another will not require readjustment of the probe compensation.

3.3   ×1/×10 Switchable probes

Another compromise is to use a switchable probe. A switch on the probe body allows selection of either ×1 or ×10 operation with all the advantages of both. However, remember here to check when taking measurements, to divide by 10 or not! according to switch position. As a general rule, always leave these probes set to the ×10 position unless ×1 is specially needed. These ×1/×10 switchable probes are probably the best choice for general-purpose oscilloscope work, but remember not to get caught out by the switch position when measuring voltage amplitudes! Also the probe bandwidth will be much lower, and the loading effect much greater on the ×1 position (see Sect. 3.1), so do not get caught out by these limitations.

The same rule applies here about resetting the compensation trimmer on the probe (×10 position) when moving the probe between channels or oscilloscopes (see Sect. 3.2).

Apart from the probes already discussed, there is a variety of special-purpose types that can be used for particular applications.

3.4   High voltage probes

Whereas general-purpose probes have an input voltage limitation of about 500 V or so, higher voltage versions in the range 1000–50 000 V are fairly commonly available (1–50 kV). Associated with this higher voltage capability is high input resistance, usually 10–1000 Mohms so the sensitivity of the oscilloscope is considerably reduced. Consequently, only relatively large a.c. signals can be observed. Because of the large amount of insulation required to make these probes safe to use, they are physically rather large and unsuitable for use in confined areas or with closely packed components. High voltage probes are generally unsuitable for high frequency use, and their upper frequency limits should be checked before use. Often they are only suitable for use up to a few megahertz or even possibly the kilohertz ranges.

3.5   Current probes

The vertical input to an oscilloscope is scaled for voltage input, so the vast majority of probes (and applications) are voltage based. However, there are several current probes now widely available which convert the alternating current flowing through a single conductor to a calibrated voltage. Although there are active current probes which require their own power supplies, only the passive type will be discussed here.

These passive current probes look very similar to the standard voltage probes, and they connect directly to the oscilloscope vertical input connector in the same way. However, there are two main differences: instead of a tip or hook connector on the end of the probe, there is a partly retractable loop of ferrite (magnetic material) to enclose the current-carrying conductor; and the input sensitivity must be scaled according to the probe sensitivity and the oscilloscope attenuator setting. Figure 3.5 shows a typical current probe, with the sliding cover open for clarity.

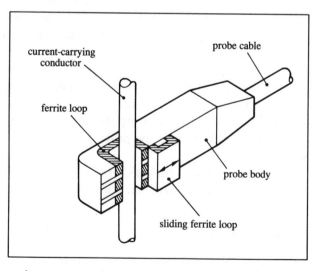

Figure 3.5. A passive current probe.

This cover is normally spring loaded into the closed position, and must be pushed back while the probe is located over the conductor carrying the current waveform to be examined. When the cover is released it closes the magnetic loop around the conductor, and this loop effectively forms the single turn primary winding of a transformer. The secondary or output from the transformer is fed down the cable to the oscilloscope input. The sensitivity of the probe must then be related to the oscilloscope input to give meaningful practical measurements in current/div.

The effective sensitivity of the probe will be:

$$S = \frac{S_p \times V}{P}$$

where:

S_p = value of current shown on probe range used, e.g. if probe = 5 mA per 10 mV, $S_p = 5$.

V = attenuator voltage range, e.g. if oscilloscope set to 5 mV/div, $V = 5$.

P = value of voltage shown on probe range used, e.g. if probe = 5 mA/10 mV, $P = 10$.

Then in the above example, current sensitivity of probe will be:

$$S = \frac{5 \times 5}{10} = 2.5 \text{ mA per division}$$

That is if the current probe gives 5 mA/10 mV, and the oscilloscope is set to 5 mV range, the probe will display 2.5 mA/div on the oscilloscope screen. (*Remember to set the Y variable control to the CALibrated position.*)

It is convenient when using current probes to set the attenuator switch on the oscilloscope to the same value shown by the probe when this is possible. Then the system is direct reading, e.g. in the above example the probe was 5 mA/10 mV, so set the oscilloscope range to 10 mV. Now the system will give 5 mA/div (5 mA being the value shown on the probe).

One slight difference in the use of current probes is the polarization of the input signal. According to which way round the current-carrying conductor is placed in the probe, the output voltage will be positive or negative. Current probes are normally marked with a positive sign (+) or an arrow. The more positive point of the circuit should be connected to the + sign on the probe, or the current flow aligned with the arrow, to produce a positive voltage at the output.

3.6   High frequency probes

Within the range of standard ×10 voltage probes, there are several types available for use at higher frequencies. 'Standard' inexpensive types have a top frequency limit in the range 100–150 MHz. More expensive models may have 250 or 350 MHz bandwidths. On these higher frequency types, there may be extra compensation adjustments which must be set with a high frequency, fast risetime pulse.

3.7   Active probes

Another type of probe available for high frequency use is the active probe. This incorporates a field effect transistor (FET) in the probe tip itself, and of course means that a d.c. power supply has to be fed to the probe tip to operate the FET.

This FET input system, right at the probe tip, forms the input amplifier to the oscilloscope, with a high impedance input and low impedance output. So the only capacitance loading at the probe input is from the FET itself and, of course, the physical probe tip. All the cable and switching capacitances are eliminated. Consequently a probe tip capacitance of 2–3 pF is possible with these probes. The low output impedance of the FET then drives the cable between probe body and oscilloscope input, enabling a high bandwidth probe system. Active probes are usually of unity gain, i.e. 1:1 voltage transfer ratio. They may have an input resistance of 1 or 10 Mohms, but this does not influence the probe attenuation since it is *not a passive divider probe*.

Inevitably these probes are more expensive than passive probes, and they have the disadvantage of requiring a d.c. power supply fed to the probe body. Sometimes this is obtained from the oscilloscope, and sometimes from a d.c. mains adaptor unit supplied with the probe.

Active probes are invaluable for investigating circuitry where high frequencies, high impedance or tuned circuits are involved, since the low 2–3 pF probe capacitance has negligible loading effect on the circuit being examined.

APPLICATION EXAMPLE

Probe use when testing transistor amplifier square wave response

In this example we shall look at the square wave response at the collector of a transistor in an amplifier circuit. The purpose of the example is to demonstrate the importance of correct probe use.

Figure 3.6 shows part of a high frequency amplifier circuit. The input circuit is not shown, but the section we are concerned with is the output stage, transistor T6.

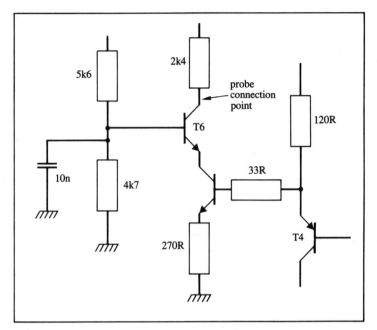

Figure 3.6. Circuit used for testing the high frequency response of a transistor amplifier using correct probes.

In order to test the response of the amplifier, we shall connect a high frequency, fast risetime square wave to the input of the amplifier. A dual trace oscilloscope is then used to examine the square wave response at the collector of T6.

The square wave input signal is supplied by a signal generator. It has a frequency of 1 MHz and a risetime of 5 ns. The square wave is connected to the amplifier input, and the signal amplitude adjusted to 20 mV p.p., which is within the operating range of the amplifier.

A ×1 probe is connected to the CH1 input of the oscilloscope, and a ×10 probe to the CH2 input. The ×1 probe has no adjustment, but the ×10 probe must first be frequency compensated to give a square wave response with a 1 kHz square wave (see Fig. 3.2).

Both the CH1 and CH2 probes are to be connected to the T6 collector, but *not at the same time*. The capacitive loading effect of the probes will change the square wave response at the point we examine, but one, the ×1 probe, will have a much greater effect.

First connect the CH1 probe to the collector of T6, as in Fig. 3.6. Adjust the CH1 attenuator to give a convenient display and set the timebase switch to 0.5 μs/div. Select CH1 trigger source on the selector switch. The upper trace of Fig. 3.7 shows the result. The front edges of the square wave show a slow rounded appearance. This is

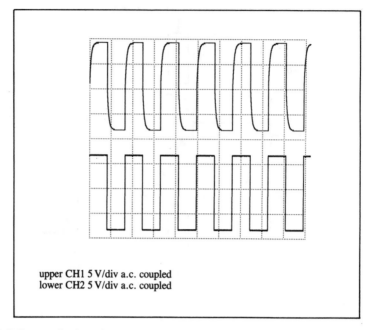

upper CH1 5 V/div a.c. coupled
lower CH2 5 V/div a.c. coupled

Figure 3.7. Screen display of squarewave response using first ×1, then ×10 probes.

due to both the slow response of the ×1 probe in itself, and the slowing of the response of the transistor T6 due to the extra capacitance of the ×1 probe (about 50 pF).

Now disconnect the CH1 probe from the circuit, and instead connect the ×10 probe from CH2 of the oscilloscope. Also select CH2 on the trigger source switch. Now observe the waveform on the screen. The result is shown as the lower trace in Fig. 3.7. The rising and falling edges are now much steeper (faster) and the corners more square. In this example the timebase variable was adjusted slightly so that the rising and falling edges of the waveform did not coincide with the graticule lines. This is now a more accurate picture of the waveform at that part of the circuit. Remember that there was no change of conditions to the signal or to the circuit. It is only a case of looking at the same point twice, but with different probes. You can conclude from these results that the lower trace picture is still not quite the true one. The ×10 probe must still be loading the circuit slightly, even with its lower capacitance and higher bandwidth. However, the probe bandwidth should be related to the bandwidth of the circuit under test. If possible, use a probe (and oscilloscope) at least 10 times the bandwidth of the circuit your are testing.

Note that the waveforms shown in Fig. 3.7 are both recorded at 5 V/div, but in each case the probe attenuation factor is included. So the Channel 1 waveform was obtained with the ×1 probe and an attenuator position of 5 V/div (5 V/div overall). The lower, Channel 2 waveform was captured with the ×10 probe and the attenuator set to 500 mV/div (5 V/div overall).

4
Trigger circuits

One of the most developed but least understood features of modern oscillo-scopes is the trigger circuit. The vital role of the trigger circuit is to 'lock' the sweep generator cycle to the input waveform so as to present a solid stable picture on the screen. Historically, oscilloscopes used to have synchronization or 'sync' circuitry to adjust the frequency of the oscilloscope timebase to a multiple or sub-multiple of the input signal frequency. That is, the sweep rate was actually changed by the input signal frequency. Since the sweep speed was variable in this way, it was not calibrated to fixed points on the timebase switch. Hence it was not possible to make time (or frequency) measurements of the display signal. A superior development was to make the timebase run at a particular sweep rate set by the TIME/DIV switch, and control its starting point relative to the input signal. That is, the speed at which the spot travels across the screen is only controlled by the timebase switch (and VARiable control) setting and *not* by the input signal frequency. The trigger circuit compares the input signal with a fixed reference voltage so that when the input signal voltage passes through a certain level, it makes the sweep generator start its cycle. The trigger signal is then prevented from influencing the timebase (inhibited) until the sweep cycle is complete. The sweep gener-ator then waits until the next trigger pulse 'fires' it again. So what are the requirements of a trigger circuit? They are to accept an input signal of any shape, size and frequency, and provide a sharp pulse output which occurs at exactly the same point on each cycle of the input waveform. There are two main modes of operation for the trigger circuit: automatic and manual.

4.1   Automatic trigger

It is probably easier to understand how the trigger works if we briefly consider the timebase operation. The trace on the oscilloscope screen is obtained by moving a spot of light from left to right at a fixed rate determined by the timebase speed control (Time/Div switch). The repeated movement of the spot makes it look like a line, the spot being rapidly returned from right to left (flyback) while blanked out. Now a feature of modern oscilloscopes is to display the trace even when no input signal is present and therefore the sweep is untriggered. This allows you to see exactly where the trace is when there is no input. Then when an input is applied, the sweep is automatically triggered by the incoming signal. This system is usually known as 'Bright Line Automatic trigger'.

Many modern oscilloscopes do have (usually limited) control of the trigger point on AUTO, the most popular system being Peak Auto trigger. However, on many oscilloscopes there is no control over the trigger point in the automatic trigger mode, the (manual) level control being disabled. This has the slight disadvantage that in the automatic mode the oscilloscope will not necessarily trigger on every type of signal that you present to it. But it has the enormous advantage for most input signals examined, that the operator has the freedom to move the input probe to different parts of the apparatus being checked without having to constantly readjust the trigger controls. So the automatic trigger mode is a way of allowing the sweep generator to continuously recycle on its own (free run) when there is no input signal applied; then to automatically trigger the timebase by an incoming signal when it is applied, without having to adjust any controls. When the timebase is triggered by the input signal, it ceases to free run as an oscillator, and only 'fires' once each time a trigger pulse is received by it. In the automatic mode, when an input signal is present, a 'detector' circuit is activated by the presence of the trigger output signals. Since these trigger output signals signify a vertical input signal to the oscilloscope, this detector circuit takes control of the sweep generator cycle and prevents it from free running. The sweep now fires once on each received trigger signal. At the end of each sweep (in the triggerable condition), the detector circuit 'waits' for the next trigger signal. If a trigger pulse arrives, it allows the sweep to be triggered. If, however, there is no trigger pulse within about 30 milliseconds (30 ms), it is taken that there are no further trigger pulses coming, and the sweep generator is returned to the free running (bright line auto) mode.

4.2 Manual trigger

The above (automatic) system can usually be overridden by switching the trigger circuit to the manual mode. In this condition, no trace (or spot) is visible on the screen until an input signal is applied. Only when an input signal occurs, sufficient to trigger the timebase, does the trace appear.

Now let us suppose that our oscilloscope is set to this manual trigger condition, with no input signal to the oscilloscope. Although the spot is not visible on the screen, its position (x and y coordinates) is determined by the X and Y amplifiers. With the X and Y position controls set to mid-range, the unseen spot will be at the vertical centre, and horizontal left edge (see Fig. 4.1).

The sweep circuit is waiting to be 'fired' by a signal from the trigger circuit.

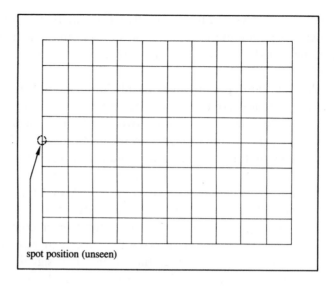

Figure 4.1. The horizontal position of the (blanked) spot prior to the sweep cycle.

When a sharp single trigger pulse occurs, the spot will immediately be unblanked and then driven at a uniform rate across the screen by the sweep generator circuit. In the manual trigger mode, the trigger level control is active and is used to select the starting point or 'trigger level' of the displayed waveform. The level control is discussed later, but basically it sets the fixed voltage reference on the trigger comparator mentioned above. But remember when the oscilloscope is in this mode, the screen display disappears as soon as the input signal is removed.

One of the most common causes of apparent 'loss of trace' on an oscilloscope is when set to manual trigger mode and with no input; or with an input, and the level control out of range of the input signal. So remember this when your instrument appears to have failed and you cannot obtain a trace on the screen.

So now let us return to the trigger circuit itself. We have seen what is required of the trigger circuit, so now how does it do it? First the signal is connected through from the Y input amplifier via some switches and filters. These usually comprise a trigger SOURCE switch and a trigger MODE switch.

4.3   Trigger source

The source switch determines where the trigger signal is taken from, for example internal or external, CH1 or CH2 in the case of a dual trace instrument, and perhaps line (mains) frequency. Here a sample is taken from one of the secondary (output) windings of the mains transformer. Obviously this sample is synchronous with the mains (line) voltage. This source selector can be used to advantage. It is not always necessary or best to trigger from the signal you are observing. With many signals, particularly complex ones, it is easier to trigger by applying another signal to (say) Channel 2 input, or the external input, to use as a trigger source. Then the signal may be displayed on Channel 1, but the trigger source selector set to Channel 2, or perhaps external. So as the function name suggests, the switch selects the source of the signal that is to trigger the sweep, irrespective of the signal that is displayed on the screen.

4.4   Trigger mode

This switch determines the coupling and filtering of the trigger signal on its way to the trigger circuit. Filters consist of frequency selective circuits which only allow certain frequency signals to pass through. For example a low pass filter will allow low frequency signals up to a certain frequency to pass through. Signals above this frequency will be blocked or reduced by the filter, and will not reach the trigger comparator. The most common mode used is a.c. coupling, which provides a stable triggering performance for the majority of simple mid-frequency waveforms. For many applications, however, a steady trace can only be obtained by the use of other trigger modes. If the waveform contains high frequency interference on the observed lower frequency signal, the LF (low frequency) filter position will allow the intended part of the signal to pass into the trigger circuit. The higher frequency interference will be blocked by the filter and thus prevent false triggering.

Similarly the HF (high frequency) mode can be used to remove the low frequency part of a signal and allow the higher frequency signal of interest to reach the trigger circuit. If the input signal frequency is very low, say less than 30 Hz, then d.c. coupling is invaluable. This mode actually d.c. couples the vertical amplifier into the trigger circuit, and overcomes the problem of the low frequency bandwidth limitation of the a.c. coupling capacitor. On some oscilloscopes, where the vertical position control is in circuit before the trigger pick-off point, the trigger point will be effectively fixed on the oscilloscope screen. So on the d.c. trigger mode, adjustment of the vertical position of the trace will appear to move the trigger point around the input waveform.

You should take extra care when using the HF and LF filter positions on different oscilloscopes. On some models the selector may be marked 'LF' on the front panel, which indicates low frequency operation. This means that it is a low frequency 'pass' filter, and will allow low frequencies through, while reducing or blocking the high frequencies. However, on other models, the panel may be marked 'LF REJ' which means low frequency reject. This actually means the low frequencies are reduced or rejected by the filter, and is thus the opposite of 'LF'. A similar situation occurs with 'HF' and 'HF REJ'. So in general

- LF is the same as HF REJ
- HF is the same as LF REJ

You will need to consult the technical specifications in your oscilloscope manual to determine the exact frequency limits for each filter position.

The signal finally reaches the trigger circuit input and can proceed in order to 'fire' the timebase. Let us suppose that the trigger input signal is a 1 kHz sinewave and we are using a.c. coupling on the trigger selector switch.

4.5   Comparator

The signal is fed into a comparator circuit which has a fast switching action, so that when the input sinewave causes the comparator to change states, its output changes levels within a few nanoseconds (1 nano-second = 1 ns = 1×10^{-9} s = one-thousand-millionth of a second). The comparator is a two input system, one input being used for the analog (1 kHz sinewave) input, and the other fed by a d.c. reference voltage (from the trigger level control). When the instantaneous level of the input signal passes the d.c. level set on the reference input of the comparator, the output immediately changes states (see Fig. 4.2). So the comparator is really the heart of the trigger circuit. It compares the incoming analog signal from the chosen verti-cal amplifier with a fixed d.c. reference voltage. When the level of the signal

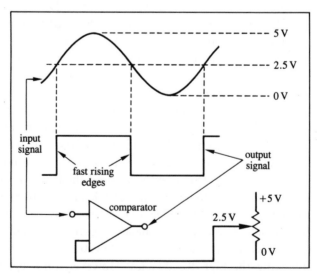

Figure 4.2. The relationship between an input sinewave signal on the CRT screen, and the trigger comparator output waveform.

voltage is equal to the reference voltage, its output very quickly changes states or voltage level.

Figure 4.2 shows a TTL type comparator using a 0 V and + 5 V supply, and the reference input set half-way at 2.5 V. When the input signal level is below the 2.5 V reference, the comparator output level is low at 0 V. As the rising signal voltage crosses the 2.5 V point, the comparator output level immediately goes high, to + 5 V. When the signal level falls again, and passes through the 2.5 V level, the comparator output immediately falls again to 0 V. If the d.c. input reference level is altered by adjusting the trigger level control, the fast edge transitions will thus occur at different points on the input waveform. Since these fast edges trigger the timebase, it can be seen that the level control thus starts the sweep on the chosen part of the input waveform.

4.6 Trigger level control

The d.c. reference input is connected to a variable potentiometer (pot), the trigger level control. This can then be used to alter the trigger point on the input waveform by varying the reference level set on the comparator input. Each time the input signal passes through the reference level, the comparator changes stages. Whatever shape or frequency the input waveform takes, the comparator always gives the same amplitude, fast edged output, only varying in frequency and mark to space ratio (ratio of the positive part of the waveform to the negative part of the waveform in one cycle).

The trigger level control is usually only effective in the manual trigger mode, and if the level control is set to a potential that is outside the range of the trigger input signal, then the analog signal will not pass through the reference level. Hence the comparator output will not switch over. The timebase will thus not be triggered and there will be no trace on the screen.

As previously stated, this is a condition always to be careful of as a common cause of total loss of trace, and the problem is often overcome on modern oscilloscopes by using the Peak Auto trigger system which is explained later.

If the trigger circuit is set to automatic trigger mode, the level control is disabled and the trigger comparator reference input is set to the mean level of the input waveform. A trigger input detector circuit then controls the timebase state. If a trigger input signal is detected by this circuit, it sets the timebase control logic to a stable non-running state, waiting to be triggered. The next output pulse edge from the trigger comparator then fires the timebase, and subsequent pulses continue to do so giving a stable picture on the screen. Should the input signal cease for more than 30 ms or so, the detector will set the timebase to a free-running state, where the timebase generator will automatically recycle at the end of each sweep and start again. This produces a bright line trace on the screen of the oscilloscope in the absence of an input signal. This is a most useful situation, enabling the trace to be positioned to any reference point on the screen with no input. This can quickly be achieved by 'grounding' the probe tip, or setting the probe to the 'REF' position. As soon as an input signal is applied, a stable trace can be obtained on the screen without the need to touch any controls.

4.7   Polarity switch +/− (slope)

It can be seen (Fig. 4.2) that when the trigger input signal crosses the threshold level set on the comparator, the output changes states. This is the case for each half-cycle of the input waveform: once on the positive going transition and once going back on the negative going transition. So two fast edge pulses are produced by the comparator for each complete cycle of the input waveform. Both these fast pulses can be used to fire the timebase. Operation of the polarity switch selects which one of these two edges is chosen. If the + polarity is chosen, then the fast pulse that is fed through to the sweep generator is the one that occurs when the positive going transition of the input signal passes through the comparator threshold. For the majority of applications, the polarity switch enables the operator to choose whether to trigger from the positive or negative going part of the displayed waveform. There is a special case, however, where the polarity switch may need slightly different use – for displaying television waveforms. In this case, rather than choosing the polarity of the trigger, it is necessary to select the polarity according to the displayed waveform, or the special TV trigger system may not work. Television signals contain video and sync (synchronization) information, as well as other components such as sound, data, etc. If the *displayed* TV waveform has positive video information, and negative going sync, then you would normally select negative TV trigger. If the sync is positive on the oscilloscope screen, and the video negative, select positive TV trigger. Check in your oscilloscope handbook to verify that TV polarity selection refers to the sync as above, and not to the video. If your oscilloscope uses the opposite method, then use the polarity switch accordingly.

4.8   Trigger point

By using the level control in combination with the polarity switch, it is thus possible to choose any point on the positive or negative going half-cycle of an input waveform. A further application of the level control is to set a level below which triggering will not occur. In a signal where some portion of the waveform is slightly larger than the rest, this can be used as the trigger point to start the sweep cycle. This is particularly useful if there is a constant

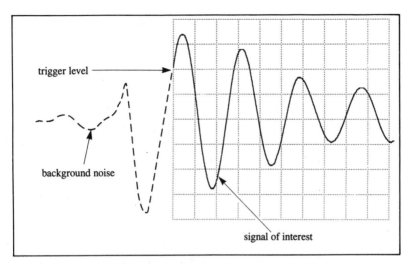

Figure 4.3. The use of trigger level threshold to start the sweep only when signal amplitude is large.

background 'noise' signal present at the point of interest (see Fig. 4.3). By using the level control in the manual trigger mode, the reference voltage can be set on the trigger comparator to be just higher than the voltage level of the background noise. Then the noise level will not fire the timebase. Only when the larger signal voltage occurs will the trigger comparator operate, and the timebase fire.

4.9   Trigger indicator lamp

Most oscilloscopes feature an indicator lamp or LED (light emitting diode) on the front panel to indicate operation of the trigger circuit. When the signal crosses the trigger comparator threshold, the lamp flashes, and for repetitive signals the lamp will flicker. At higher frequencies the lamp appears to be continuously lit.

For most medium and high frequency repetitive signals, the trigger lamp serves no great purpose as the display on the oscilloscope screen is the best indicator itself, actually displaying the trigger point as the start of the waveform. However, there are some circumstances where the indicator lamp is extremely useful. When you are trying to trigger on signals with a very low repetition rate, the lamp may be the only indication of whether the timebase fired or not. These may be fast edged signals which occur very infrequently, or just very low frequency repetitive signals. By carefully adjusting the level control, and watching the trigger lamp, usually in the d.c. trigger coupling mode, you can tell when you have set the correct trigger point by the repeated flashing of the lamp. Then you can concentrate on selecting the best timebase and brightness conditions to display the signal.

4.9.1 *Manual trigger*

When using the manual trigger mode, the display will only occur on the screen when the oscilloscope is triggered. Adjusting the level control will determine the trigger point on the input signal. However, with the level control out of range, the screen is blank. This means that until you connect your probe to a signal to trigger the timebase, the trace is effectively 'lost'. So when you have completed your application using manual trigger mode, switch back to the AUTOmatic trigger mode. Then, next time you use the oscilloscope you will start off with a visible trace. If you are using manual trigger and if adjustment of the level control does not produce a display, look at the trigger indicator lamp. If you can adjust the level to get the lamp continuously lit (or flashing with a low frequency input signal), then set it to the triggering position (lit). Now find out why there is no display. Usually adjustment of the timebase and/or attenuator controls will produce a display. Also try the vertical and horizontal position controls. Watch the trigger lamp while you take it below the triggerable range, and further careful adjustment of the level control may be necessary to relight the trigger lamp.

4.10   Trigger mode and source

Although the screen may have a large picture displayed on it, you may sometimes find it impossible to trigger on any part of the waveform or get a sensible display. A check of the trigger lamp may show that the input waveform is not reaching the trigger circuit (trigger lamp not lit). First set the trigger circuit to AUTOmatic trigger mode. Then check the path from your input signal to the trigger circuit.

1 Check that the trigger channel selector corresponds to the signal input channel. If the signal is connected to CH1 Y input, select CH1 trigger.
2 Check that the trigger is set to INTERNAL or EXTERNAL as required. If the signal goes in the CH1 Y input, select INTernal trigger.
3 Check that the coupling is appropriate to the type of input signal. For most medium frequency signals select AC trigger, for low frequency signals select LF or DC coupling, etc.
4 Check the other trigger selections, TV trigger, ALTernate trigger, etc., and set to NORMAL or OFF.

Once these checks are carried out, it should now be possible to get the trigger lamp to light, and adjustment of the timebase switch should produce a sensible display, except in the case of complex waveforms.

4.11   Alternate trigger

This is a special facility which enables the oscilloscope to alternately trigger from two unrelated signals. In general, a dual trace oscilloscope is used to display two different signals, one on Channel 1 and one on Channel 2. The two signals may be before and after amplification for instance, in which case it is the same signal but from different points in a circuit. Or the two signals may be (say) a sinewave and square wave in the same function generator. In either case here and in general, the Channel 1 and Channel 2 signals must be 'time locked'. That is, they must be generated from or synchronized with the same time reference. If this is the case, then the two signals can be presented as a stable display on the screen. However, if the signals are not 'time locked', then one of them will 'run through' on the screen. For instance if you use two separate signal generators, with both their outputs set to (say) 1 kHz, 1 V p.p. sinewaves, and apply these signals to the Channel 1 and Channel 2 inputs, only one will lock. If you select Channel 1 trigger, then Channel 1 display will be stable. The Channel 2 sinewave will slowly run through, and vice versa if Channel 2 trigger is selected.

Now if you select ALTernate trigger both signals can be locked. The ALT trigger function must be used in conjunction with the ALTernate vertical display mode, *not* chop. In this alternate vertical mode, Channel 1 and Channel 2 signals are alternatively displayed on successive sweeps of the timebase.

Now with ALTernate trigger selected, when the Channel 1 vertical signal is displayed, it is also routed to the trigger circuit to fire the timebase. At the end of that sweep, trigger control passes to Channel 2. On the next sweep, triggered by the Channel 2 input signal, the Channel 2 vertical signal is displayed. This process then repeats with the Channel 1 signal triggering the timebase, and Channel 1 displayed; then Channel 2 firing the timebase and Channel 2 displayed, and so on. Thus for fast sweep speeds above about 1 ms/div, the two traces appear simultaneously as stable triggered displays.

There are often limitations to the use of this trigger mode. Usually:

1   ALTernate vertical display mode must be selected.
2   Both traces must be within the screen area.
3   Manual trigger level control operation may be necessary.
4   The two traces may need to overlap on the screen.

However, these limitations are quite practical, and this facility then allows signals to be compared from completely different time sources. It is especially helpful as a faultfinding facility since the signal from a faulty instrument can be compared with the equivalent signal from another instrument of the same type. After using the ALTernate trigger facility, do not forget to switch it off as it will make ordinary triggering seem erratic.

4.12   Complex waveforms

Although the trigger lamp may be lit it is sometimes impossible to obtain a sensible waveform display. It may be that complex signals such as digital word sequences, television waveforms, modulated carriers, etc., require some special processing by the trigger circuit to isolate the part of the waveform of particular interest.

If the nature of the waveform is known, it is possible to discriminate the required part by use of say LF or HF coupling to eliminate high or low frequency content from the signal reaching the trigger input stage. In the case of TV waveforms, special processing circuits in the scope can be used. These are sync separator circuits which extract the television line and field synchronizing pulses from the composite waveform.

When these facilities are employed, watch the trigger lamp and use the level control on manual trigger mode to show when the signal is correctly triggered. And, of course, always watch the screen display.

4.13   Low frequency signals

For triggering very low frequency signals less than 30 Hz, it is necessary to use the LF or even d.c. coupling of the trigger mode switch, and manual trigger. Since there is no trace in manual mode until a trigger input is present, the trigger lamp will indicate when the level control is in the correct position. When the timebase is set to a very slow speed, and the display appears as a moving dot of light rather than a continuous waveform, the trigger lamp may be the only indication of the correct trigger point on the waveform. Some patience is often required while slowly adjusting the level control and timebase switch, while observing the trigger lamp and the screen display.

4.14   TV trigger

Many modern oscilloscopes are equipped with a special trigger function specifically for television waveforms. The problem with trying to trigger on TV waveforms arises from the fact that the composite waveform received in a TV set contains all the information in one complex signal. The video information: colour, picture content, etc., is superimposed on the sync information: line and field synchronization pulses. In order to display any one of these signals on an oscilloscope, without triggering on them all simultaneously, a special circuit is provided to sort them out. Although specialized oscilloscopes can isolate any part of these waveforms to look at (say) one particular TV line, general-purpose scopes usually have the facility for line sync or field sync only. In this case it is not possible to select which line (1–625) or field (1 or 2, odd or even) the scope will trigger on. However, it is still

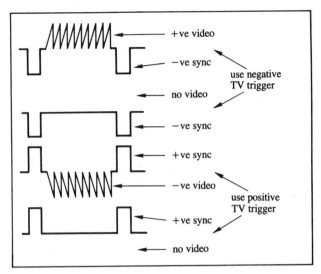

Figure 4.4. The polarity of TV line pulses to identify positive and negative video (negative and positive sync).

a very useful function and the TV trigger circuit takes two forms: active sync separator and passive sync separator. In the latter case, the passive version is a high pass (TV line) or low pass (TV field) filter to discriminate between the low frequency field sync pulses and the higher frequency line sync pulses. The active separator employs semiconductors to process the composite signal and extract the line and field pulses from the complex waveform.

In the case of TV trigger circuits, the polarity switch is particularly important. When used in conjunction with a TV sync separator it may have a different function altogether. In this case, instead of using the polarity switch in the normal way, to display the waveform edge you wish to trigger from, the switch *must* be set to match the polarity of the sync at the point being observed. If the video is positive going, and the sync tips negative (see Fig. 4.4), then select NEGATIVE TV trigger. If the video is negative and the sync tips positive going, choose POSITIVE TV trigger.

Even if no video information is present on the signal, the correct polarity must still be chosen (Fig. 4.4).

4.15   Television receivers

It is worth noting here that television receivers may have a live chassis and must be connected through an isolating transformer before work can commence, in order that the chassis can be connected to ground by the oscilloscope probe ground. In normal use, a television chassis may be live; being connected directly to the mains supply, so *no attempt should be made to work on a television receiver by untrained personnel. No attempt should be made to work on a television without special isolating or safety equipment.*

4.16   External trigger

All references to triggering so far have assumed that the trigger signal is derived from the same signal entering one of the vertical input sockets. So the same signal that is displayed on the screen is actually fed through (via conditioning circuitry) to the trigger input system.

Now most oscilloscopes also provide a front or rear panel socket for external triggering. Here a signal may be connected via this (external trigger) socket directly to the trigger input circuit. The applied external signal need not be the same size, shape or frequency as the vertical input signal, but the one essential condition is that it is synchronous or 'time locked' to the displayed signal. In Fig. 4.5, two different signals are displayed on CH1 and CH2 of the oscilloscope. The trigger circuit is set to EXT trigger mode and the signal applied to the external trigger socket is a spike signal synchronous with the displayed waveform, but different in shape and frequency, as shown.

As can be seen in Fig. 4.5, two signals, both different from the external trigger signal, can be solidly displayed on the screen, but this can only happen if the external trigger signal is derived from the same source as the other two (displayed) signals. *The displayed signals and the external trigger signal must all be connected to the same clock, oscillator, or timing reference.*

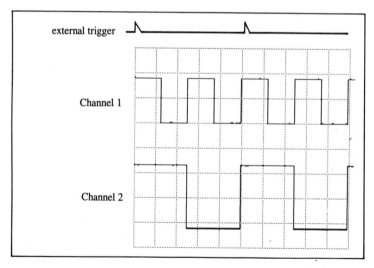

Figure 4.5. The time relationship between an external synchronous trigger signal, and two digital pulse waveforms to be triggered.

It is no use trying to external trigger with a signal from a different timing reference, even if it is exactly the same frequency, or shape or size (or all three) as a displayed signal.

4.16.1 Use of external trigger – 'hands off triggering'

When checking through circuitry with a scope probe, perhaps following a signal path, transferring the probe tip from one point to another may mean constantly resetting the trigger controls, specially the trigger level. This is due to the variation in size and shape of the signal at each point observed.

If the trigger circuit is set to the AUTO mode, when the displayed signal amplitude drops below the trigger threshold and is thus too small to trigger the timebase, the sweep will 'free run' and give an unsynchronized display. When the trigger circuit is used in the manual mode, each time the probe is unconnected, there is no trace on the screen; when the probe is connected the level will probably need adjusting to give the best trigger setting for the new input signal. Now if the oscilloscope is externally triggered from a point in your circuit which is time locked to (synchronous with) the signal you are tracing, then the probe can be freely moved around the circuit without any need to readjust the trigger. Simply connecting a probe from the external trigger input to one point in your circuit where there is a strong signal may be all that is needed. Connect a probe from the strong signal point to the external trigger socket, and select EXTernal trigger on the trigger source switch. Also connect the CH1 (or whichever channel you prefer to use) probe to the same point. Now, while externally triggered, adjust the trigger level and filter controls as necessary to solidly lock the CH1 trace on the screen. Once the trigger controls have been set, you can now freely move the CH1 probe around the circuit you are testing, and the timebase will remain solidly triggered by the external trigger probe connection. Note that with dual trace oscilloscopes, the same benefit can usually be achieved by using one vertical input as the trigger input, constantly connected to the trigger source signal, while the other channel is used to 'probe around' the circuit you are checking.

4.16.2 Signal isolation

With many complex signals, such as a television composite waveform and digital signals, the combination of different pulse widths and amplitudes within one period of the signal makes it difficult to trigger. Since most circuits that process these types of signal usually simplify and extract specific elements from these complex waveforms, it is useful to go down the signal processing chain, and use the *final* signal as the external trigger source for the scope. Once having locked to this signal, on external input, use the CH1 and CH2 probes to go through the preceding circuits. Now although the previous waveforms may be far more complex, the use of external trigger isolates the signal of interest from the main complex waveform.

Again it should be noted that the above techniques for external trigger can equally be achieved on dual trace oscilloscopes by the use of the second vertical channel as the trigger source, while the first channel is used for signal tracing. However, on some oscilloscopes, the second vertical channel can only be used as the trigger source if it is actually displayed on the screen. In this case the constant presence of a waveform on the screen overlapping and confusing the signal you want to follow, causes problems. This is overcome by the use of external triggering as detailed above. When using the external trigger facility, carefully check the requirements of the signal applied to the external trigger socket. Check the minimum voltage required for trigger and maximum voltage input without damage. Also remember that the larger the signal applied to the external socket, the less effect the level control will have to set the trigger point. A slight disadvantage of this external triggering method is that you do not know exactly what you are triggering on. So given the choice of external or Channel 2 as the triggering channel, you may prefer to use Channel 2 as you can then see the trigger point on the screen (when you display Channel 2).

4.17   Peak auto trigger

This special kind of trigger function has been left to the end of the trigger chapter since its operation contradicts other facilities already mentioned. It combines the advantages of auto trigger with the use of the level control. That is, it provides a trace on the screen when there is no input signal (bright line auto) and yet allows the level control to operate to some degree to select the trigger point. Normally the level control is only operational in the manual trigger mode. When the level control is altered, it can select a trigger point anywhere between the top and bottom peaks of the input signal. Also at each end of the level control range, it can be set outside the limits of the input signal variation. Then the input signal will not cross the comparator threshold, and no trigger pulse is produced. However, with peak auto trigger, the level cannot be set beyond the range of the input signal, so the display cannot be 'lost'.

The principle of the peak trigger system is that the voltage range of the level control is proportional to (and may be actually derived from) the size of the input signal to the trigger comparator.

The range of the level control is preset so that it is just less than the size of the trigger input signal, say 75 per cent. For example, say that the signal on the screen was a sinewave of 8 cm peak to peak. The range of the level control

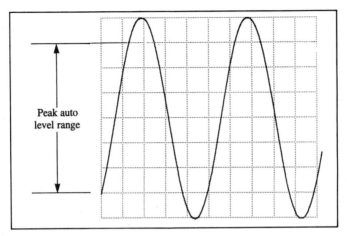

Peak auto
level range

Figure 4.6. The range of trigger level control on the displayed signal using the peak auto trigger system.

would be 75 per cent of $8 = 6$ cm. So the level range would extend up and down 3 cm from the centre crossover point (see Fig. 4.6).

Now if the signal input was reduced to 1 cm p.p., the level range would only be 75 per cent of 1 cm $= 0.75$ cm. So now with the level control set at one end, the small signal still falls within the level control range.

Normally, the auto function is disabled on the manual trigger position, so now the level control gives a wide range greater than the screen size as before (see Sect. 4.2).

The peak auto trigger mode is a superb system for most applications of the oscilloscope. It allows some degree of control of the trigger point of the signal, while still maintaining the 'bright line' trace on the screen when the probe is moved and there is no input signal. Of course there is the disadvantage that you cannot trigger over the whole amplitude range of the displayed signal, so in this case, and with many complex waveforms, normal manual trigger will still be necessary. Although the above example shows the level range limited to 75 per cent of the signal amplitude, modern auto trigger systems may allow almost complete level control to the positive and negative peaks of the signal.

APPLICATION EXAMPLE

Use of dual trace as differential amplifier

The object of this example is to use the dual trace system as a differential amplifier to observe distortion in an amplifier. The distortion is

Hands-on guide to oscilloscopes

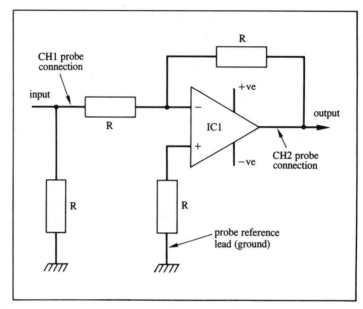

Figure 4.7. Circuit diagram of an operational amplifier system showing the probe connection points for signal monitoring.

caused by overloading the input of an operational amplifier (op-amp). That is, when the input signal amplitude reaches a certain level, the amplitude output from the op-amp cannot increase further due to being limited by its own maximum ratings, and the subsequent supply rails connected to it. The relevant part of the circuit diagram is shown in Fig. 4.7.

A signal generator is used to supply a sinewave input to the system we are studying. A ×10 probe is connected between the CH1 oscilloscope input and the op-amp input point as shown in Fig. 4.7. A second ×10 probe is connected from the CH2 input to the op-amp output point as in Fig. 4.7. Both input coupling switches are set to a.c. While observing the CH2 output waveform, the signal generator amplitude is slowly increased until distortion occurs on the sinewave display, and the amplitude is noted. Now reduce the signal generator amplitude to half its noted level causing distortion, to ensure that this level of signal is well within the handling range of the op-amp.

Next adjust both the CH1 and CH2 attenuators so that both the displayed traces have an amplitude of about three to four divisions, with both the VARiable controls set to the CAL position. Using the vertical mode switch, select the ADD mode and observe the resultant waveform. It should be small or even a straight line. In this example, the input signal is fed into the inverting input terminal of the op-amp, so the input and output are antiphase. If the non-inverting input was used, the in phase signals would add and result in a large display of six to eight divisions on the screen. In that case, use the INVERT function on CH1 *or* CH2, and this will result in a small difference signal as required.

In order to get both the display waveforms exactly the same size, try the CH1 VARiable control, and adjust carefully to give a straight line on the screen. If the display gets bigger, reset the CH1 VARiable to CAL, and try the CH2 VARiable. One or other controls should give a straight line display. When this is achieved, leave all controls set, and now increase the signal amplitude from the sinewave generator. Eventually when the signal amplitude reaches the maximum level that the op-amp can handle, distortion occurs, and produces a distortion waveform as shown in Fig. 4.8.

Careful up and down adjustment of the signal generator output will find the exact point where distortion occurs. Once you find this point, leave all controls set, and disable the ADD mode, reverting to the normal dual trace mode. Now set both the VARiables to CAL.

If the waveforms are now too large for the screen, turn the attenuators counterclockwise until both traces are suitably displayed. Figure 4.9 shows the typical result of this exercise. The CH1 trace now shows the maximum amplitude input signal that the op-amp can

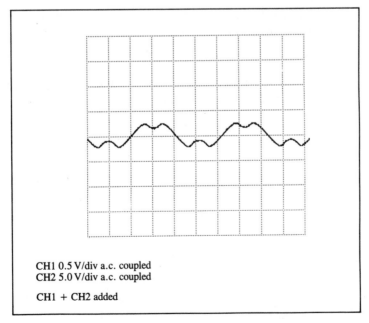

CH1 0.5 V/div a.c. coupled
CH2 5.0 V/div a.c. coupled

CH1 + CH2 added

Figure 4.8. Screen display of sinewave distortion signal showing point of overloading.

60

Hands-on guide to oscilloscopes

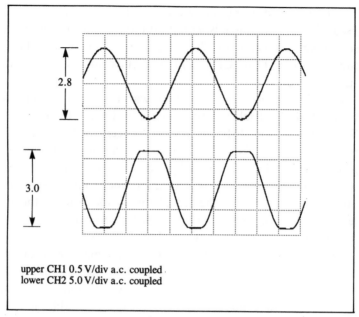

upper CH1 0.5 V/div a.c. coupled
lower CH2 5.0 V/div a.c. coupled

Figure 4.9. Screen display of maximum input and output signal amplitudes at the point of overloading.

handle. This can be measured from the graticule and in this example is 2.8 divisions.

The CH1 sensitivity at the probe tip (allowing for the ×10 probe factor) is 0.5 V/div. So the maximum input is:

$$2.8 \times 0.5 = 1.4\,\text{V p.p.}$$

CH2 shows the maximum output, and is 3.0 divisions, at a probe tip sensitivity of 5 V/div. Hence the maximum output is:

$$3.0 \times 5 = 15\,\text{V p.p.}$$

5
Timebase generator

The purpose of a timebase generator is to provide a voltage waveform to drive the spot from left to right across the oscilloscope screen, and return it back to the left-hand side. The 'sweep' from left to right is visible, while the 'flyback' from right to left is unseen, the cathode-ray tube (CRT) being blanked during this period. The sweep must be at a fixed rate and linear. This means that the speed at which the spot travels from left to right must accord with the front panel control setting and the speed must be constant across the whole screen width.

It is called a timebase generator because on a graph of voltage against time, the *time* axis is the *baseline*; while the voltage is the vertical axis. To put it another way, the voltage deviation is plotted against the timebase. The timebase waveform rate (time per division) and duration are only determined by the panel controls; the TIME/DIV switch and the time VARiable control. They are not affected by the input signal or the trigger circuit. The vertical and trigger circuits only determine the point when the sweep starts. Once started, the sweep is independent of outside influence.

5.1   Time/div switch

The control used to determine the sweep speed is the time/div switch or time/cm switch. On most oscilloscopes each horizontal major division on the graticule is 1 cm, so the two terms mean the same. This control is a multi-position switch giving a wide range of timebase speed settings. As the name suggests, the switch sets the time taken for the spot to travel 1 cm left to right on the screen. So on a normal 10 cm wide screen, the total sweep duration is 10 times the time/div switch setting. The sweep speeds are divided into three ranges:

- seconds (s)
- milliseconds (ms), thousandths
- microseconds (μs), millionths

For instance, if the time/div switch is set to 1 ms/div, the spot will travel at exactly that speed over the whole screen, giving a total sweep time of 10 ms.

The switch operates in a 1, 2, 5 sequence, so every three positions of the switch give a 10-fold speed increase. In order to make accurate measurements, it is essential that all controls are set to their CALibration (cal) position.

5.2   Variable time control

In order to 'fill in' the time ranges between any two timebase switch positions, a VARiable control is usually provided. This control has a detent or 'click' position at one end of its travel (CAL position) at which the oscilloscope is precisely calibrated. The VARiable must be set to this position before making any measurements on the screen.

The VARiable provides at least 2.5 : 1 time variation to cover the two widest ranges on the timebase switch; between the 2 and 5 positions.

The main purpose of this VARiable control is to allow an exact amount of waveform to be displayed, by adjusting until the required signal fills the screen. You can use the Time/Div switch and timebase VARiable controls as coarse and fine adjustments. The TIME/DIV switch is used to set approximately the required number of cycles of the waveform on the screen; the VARiable control is then adjusted for a more precise setting.

It can also be used as a triggering aid, however. Some complex waveforms will cause jitter when triggered. That is, there is a recurrent horizontal movement of the triggering point on the waveform. This can often be overcome by slight adjustment of the time VARiable control. This has the effect of extending (in time) the end of the sweep until just after the erroneous triggering point, allowing the desired trigger point to consistently trigger the sweep.

In the dual trace mode, with alternate display of Channel 1 and Channel 2 signals (see Chapter 2, Vertical amplifiers), each vertical channel is displayed on the screen on successive alternate sweeps of the timebase. So although there appear to be two traces on the screen, in fact first Channel 1 is displayed during the sweep, then on the next sweep Channel 2 is displayed, then on the next sweep Channel 1 is displayed, and so on.

Now with some complex waveforms displayed on Channel 1 and Channel 2, it is possible to display these signals more clearly (and separately) by slight adjustment of the timebase VARiable control. As above with the jitter problem, use of the VARiable control extends (in time) the duration of the overall sweep cycle. Thus the point at which the next sweep can start is delayed slightly, and this can often be used to advantage to 'miss' a part of the waveform from the trigger circuit, and only 'enable' the sweep cycle after a particular part of the waveform. 'Enable' in this sense means 'allow the sweep to become triggerable'.

5.3   Position control

The horizontal position of the trace can be adjusted by this control which usually allows each end of the trace to be moved past the screen centre. This control has no effect on the sweep rate, only its horizontal position. It usually has a range of just more than one screen width. So for a screen 10 cm wide, the range would be just over 10 cm, say 12 cm. This allows each end of the trace to be shifted past screen centre. That is, the left end of the trace can be shifted to the right past the centre line; and the right end of the trace can be shifted left past the screen centre. Some models feature coarse and fine horizontal position controls. This is particularly useful when the horizontal magnifier (if fitted) is in use. When the ×5 magnifier (or ×10 as fitted) is in operation, the sensitivity of the position control increases by the same factor, so the FINE control can then be used for more precise adjustment.

5.4   Timebase sawtooth generator

The timebase generator produces a waveform shaped like, and known as, a sawtooth (Fig. 5.1). This sawtooth signal is amplified and fed to the X plates of the CRT to drive the spot horizontally across the screen. It can be seen that the sawtooth waveform is divided into four time segments. When the Y input signal crosses the trigger threshold, the fast pulse from the trigger comparator 'fires' the timebase.

     The voltage begins to rise from its start level at a fixed rate until it reaches its end level. The circuitry is arranged so that this voltage ramp is equal to just

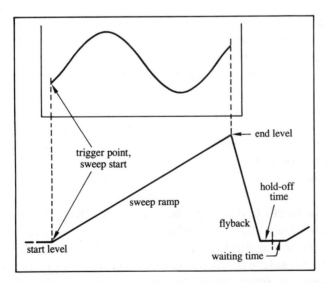

Figure 5.1. The time relationship between the displayed screen signal and the sweep ramp.

over 10 divisions of horizontal scan. As soon as the sweep ramp reaches its end level, the spot is blanked out on the screen so that the flyback is not visible. The flyback is the rapid change of the timebase waveform from the end level back to the start level. This corresponds with the (unseen) movement of the spot from right- to left-hand side of the screen, and occurs about 10 times faster than the sweep ramp. Once the voltage has returned to the original start level, another part of the circuit prevents the sweep from being fired again for a short period. This is known as the hold-off time. This is to allow the sweep ramp and flyback circuitry to stabilize into a steady state. At the end of the hold-off period, although it is not apparent from the ramp waveform diagram, the timebase can be retriggered, and a 'waiting time' then occurs until the input waveform crosses the trigger threshold again. That is, the waiting time is from when the timebase is allowed to be triggered, until the time when the next trigger pulse actually occurs. When the sweep is allowed to be triggered, it is said to be 'enabled'. The resultant trigger pulse fires the sweep and the whole process repeats itself.

In the absence of an input signal, if the trigger circuit is set to the 'automatic' mode, the ramp will begin immediately after the end of the hold-off period. The sweep generator behaves as a free running oscillator, beginning a new sweep cycle automatically at the end of each previous cycle.

If the trigger is set to the 'manual' mode, the sweep signal remains at the start level with the spot blanked. The ramp will not start to climb until a trigger pulse occurs. So in the automatic mode, there is no waiting time. As soon as the sweep is enabled, it immediately recycles. In the manual mode, the waiting time occurs between the sweep being enabled, and the next trigger pulse.

The ramp or sawtooth waveform generated to scan the CRT is often available as an external output signal. This can usually be found at the rear of the instrument indicated by a ⌁ symbol and can be used to drive other equipment in synchronism with the sweep cycle of the oscilloscope.

5.5   X Magnifier control

Most oscilloscopes are equipped with a magnifier control, either ×2, ×5 or ×10 magnification. This magnifier actually operates in the oscilloscope X output amplifier circuit. As a result, this control has two effects: it increases the effective sweep speed by the stated magnification factor; and also multiplies the trace length by the same factor. So the effect of using (say) a ×5 magnifier is to obtain a trace 50 cm long. Since only 10 cm can be viewed at a time, the X position control must be used to find the part of the waveform required. The advantage is, of course, to see a waveform in more detail in the horizontal axis. If fitted, the FINE position control should be used in conjunction with the X magnifier, for accurate control of the horizontal trace position. The disadvantage is that the trace becomes dimmer in the magnified condition, and if there is no FINE position control, the normal position adjustment is coarse and may be difficult to set accurately.

5.6 Single shot and reset

The single shot or single sweep function is a system to prevent retrigger of the timebase after one cycle of the sweep generator. Upon input of a vertical signal causing a trigger pulse, the sweep fires once then stops. It cannot be fired again until the reset button has been pressed. It usually takes the form of a bistable circuit; that is, a circuit with two stable states – 'ready' or 'armed', and 'fired'. The circuit has a similar effect to the hold-off circuitry already mentioned. At the end of the sweep cycle the single sweep circuit 'clamps' the ramp at the start level and inhibits incoming trigger pulses from firing the sweep. When the reset button is operated, the single shot bistable reverses states and allows one sweep cycle to occur when a trigger signal is received. The single sweep cycle itself then switches the single shot circuit back again so that further sweep cycles are prevented. The single shot mode is very useful when used in conjunction with a camera. Once the camera is mounted on the oscilloscope, and the reset button operated to 'arm' the sweep, the camera shutter can be opened and left. At a later time, the required signal will enter the scope input, 'fire' the timebase and be recorded by the camera. The sweep will not now operate again before being rearmed, so the open camera shutter causes no problem. At any time later, the operator can close the shutter and repeat the whole procedure if required. The single shot system cannot be reset while there is an input signal present to fire the timebase, so the input may have to be grounded or the circuit being examined switched off, to enable the circuit to arm. Without using a camera to record the single sweep when it occurs, it may be impossible to see any waveform detail. On slow sweep speeds, it may be possible to watch the waveform traced out, but since it occurs just once, it is hard to retain the detail in the mind. At fast sweep speeds, the trace is probably too fast and too dim (on a single sweep) to see anyway. However, when single shot mode is used in conjunction with digital storage, a completely different situation arises. Now the waveform is held in a digital memory, and can be displayed repetitively to appear as a solid trace (see Chapter 11, Digital storage oscilloscopes).

5.7 Variable hold-off

Looking at the sweep generator waveform in Fig. 5.1, it can be seen that after the sweep ramp, then the flyback, there is then a hold-off period. This hold-off is to allow all sweep generator conditions to reset fully before another trigger pulse is allowed to restart the cycle. Now this hold-off time can be exploited by the use of a variable hold-off control. This is a front panel control which expands the hold-off time from its normal value (about one-tenth of the sweep time) by a factor of 10, to equal the sweep time (see Fig. 5.2).

The main use of the variable hold-off control is to prevent unwanted retriggering of the timebase. If a waveform contains an irregular pattern of pulses such as shown in Fig. 5.3, then the trigger circuit will produce output pulses at all the points shown. This is because every time the input signal crosses the level set on the trigger comparator, the trigger circuit, still being operational, produces an output. (These pulses do not get through to the timebase as they are inhibited during the sweep cycle.) Now if the timebase range is chosen to display one group of these pulses, the screen will be confused with overlapping signals. This is because after the group of three pulses has been displayed, the next time the sweep fires will be at the start of the group of four pulses. Then the next time the group of three pulses, and so on. So each successive sweep of the timebase will display three then four, then three then four pulses, etc. Now these pulses may have slightly different widths, and so a confusing picture of seven intermingled pulses is presented simultaneously on the screen (Fig. 5.3).

Now by using the variable hold-off control, the sweep can be delayed from retriggering until after the next group of pulses has occurred, and then allowed to trigger on the next identical group (see Fig. 5.4). Even though the

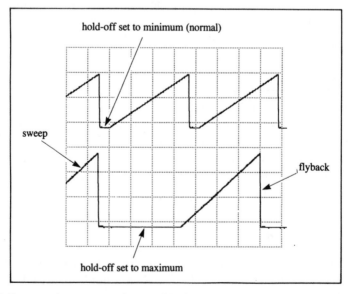

Figure 5.2. Sweep generator waveforms with the variable hold-off control set to minimum and maximum.

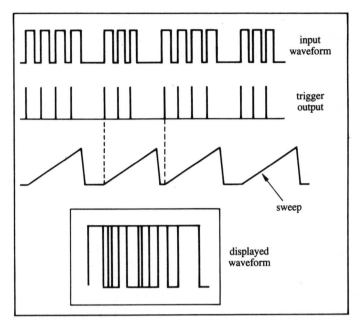

Figure 5.3. The relationship between input pulse groups, and consequent trigger, sweep and display.

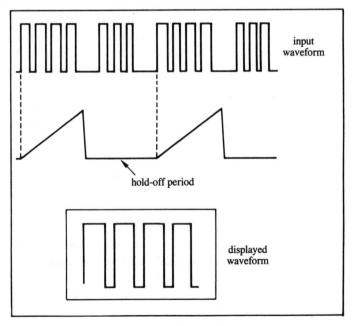

Figure 5.4. The use of extended variable hold-off to display separately the same pulse groups as in Fig. 5.3.

trigger circuit is still producing pulses coincident with every waveform transition, these alternating groups of pulses can be inhibited from retriggering the timebase because increasing the hold-off time effectively increases the overall sweep cycle time and thus extends it past the next group of trigger pulses. That is, it delays retrigger until the second group of pulses occurs and this is the same one that was already just displayed. When these overlapping waveforms occur, the easiest way to set the hold-off is by trial and error, progressively increasing the hold-off time until a satisfactory result is obtained on the screen.

After use, remember to return the hold-off control to the CALibrated or ×1 condition, to avoid unnecessary flicker of the display, or reduction in brightness.

5.8   Unblanking

The unblanking or 'bright up' circuit is the means of making the display visible (only) during the sweep cycle. Unlike the vertical, trigger and timebase functions, the unblanking system is not a 'front panel' or external facility. It is concerned with the internal circuitry only. It can be seen in the sweep generator system that there are four parts to the sweep cycle: sweep ramp, flyback, hold-off and waiting time. Of these four parts, only the sweep ramp period is made visible. A bright up pulse is derived from the sweep generator logic circuit which is synchronous with the sweep period, and this is fed to the CRT circuit. The cathode and grid of the cathode-ray tube are normally adjusted so that the beam is cut off, but when the bright up pulse drives the grid or cathode into conduction, the trace appears on the screen. At the same instant that the sweep starts its scan left to right across the screen, a very fast step pulse is fed from the timebase circuit to the CRT circuit. This fast pulse turns the CRT beam on very quickly, usually within about 20 ns. The beam (or trace) thus remains visible during the sweep cycle until the spot reaches the right of the screen. When the sweep ends, again a fast pulse is fed to the CRT, this time of opposite polarity. This causes the beam to be quickly cut off, and hence the trace is blanked. The trace then remains unseen, as the blanking pulse remains at the cut-off level, during the flyback, hold-off and waiting time periods.

As well as the bright up of the sweep period, the circuit controls other functions such as unblanking of the spot in X–Y mode, blanking of the (between channel) transitions in vertical chop mode and Z modulation. Normally when the sweep is not active, the spot is blanked, but in the X–Y mode, even though the sweep is not running, the spot must be made visible. This is achieved by switching into circuit an override signal on operation of the X–Y switch, causing the unblanking pulse to set to its 'bright' level.

In the chop mode, the spot makes many vertical transitions as the display

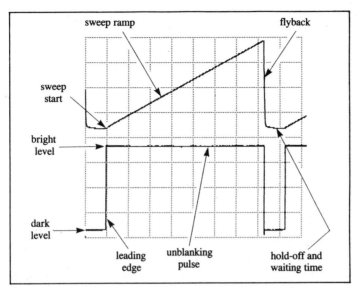

Figure 5.5. The time relationship between the sweep cycle waveform and the unblanking pulse.

quickly switches between Channel 1 and Channel 2 vertical inputs. Hence in the chop mode these switching transitions occur many times during one sweep cycle, but the movement of the spot from Channel 1 to the Channel 2 display must not be seen. A signal derived from the vertical amplifier is fed to the unblanking circuit to suppress the beam during these interchannel transitions.

Figure 5.5 shows the timing relationship between the sweep waveform and the unblanking or bright up pulse. As you can see, there are four separate time periods in the sweep waveform: the sweep ramp, the flyback, the hold-off period and the waiting time. The unblanking pulse is synchronous with the sweep ramp only, so the flyback, hold-off and waiting time are all non-visible. The unblanking pulse itself has a finite risetime on its leading edge, coincident with the sweep start, which accounts for the slight loss of trace at the beginning of the sweep on the left-hand side of the screen. The sweep only becomes visible when the unblanking pulse has reached its full amplitude and the CRT beam is turned on. This is one of the reasons for the use of a signal delay line as detailed in Chapter 2.

5.9   Z Modulation

Z mod on an oscilloscope is the variation of the brightness of the display in sympathy with an external input signal. It is like an analog signal continually adjusting the intensity control up and down. The Z mod input signal is usually connected via a socket on the rear panel of the oscilloscope, and is coupled to either the grid or the cathode of the CRT. The Z mod system takes two forms, analog and digital (TTL). With the analog type, the Z signal continually varies the brightness level of the screen display as the Z signal amplitude rises and falls, so the trace gradually gets brighter and dimmer as it travels across the screen. With the TTL type, the display is either bright or dark according to whether the Z input is high (+5 V) or low (0 V), with no intermediate brightness levels or tones.

5.10   Delay timebase

Many higher specification oscilloscopes feature a second timebase, known as the delay timebase. Systems with this facility are known by various names such as dual timebase, 'B' timebase, second timebase, delay timebase, and so on. The purpose of this second timebase is to display part of a waveform at a much faster sweep speed for closer examination. The main or 'A' sweep runs exactly as described above, as the primary sawtooth ramp generator. Then an extra circuit is required to provide a delay time. This may take the form of a monostable timer circuit with a variable delay time chosen by front panel settings. When the main A timebase starts its sweep, the monostable timing sequence is also started. The monostable time delay is less than the total A sweep duration. At the end of the monostable time delay, a signal is fed from this circuit to the 'B' or delayed timebase. The B sweep then starts its cycle, and runs faster than the A sweep.

Another method to produce the timebase delay system is to use a comparator circuit, just like the one used as a trigger comparator. Again the two comparator inputs are a d.c. voltage and an analog signal. The d.c. voltage is obtained from a multiturn potentiometer, the Delay Time Multiplier. This is a front panel control. The analog input is the sweep ramp from the A timebase. So when the A sweep starts, it is at a low voltage (start level) and is below the comparator threshold. When the A sweep ramp has risen to the voltage set on the comparator by the Delay control, the comparator output changes states, and enables the B sweep. If there is an after delay trigger system in use, the B sweep starts on the next (B) trigger pulse. Otherwise, the B sweep fires immediately. Adjustment of the Delay Time Multiplier thus varies the B sweep start position across the A sweep on the screen in the A INTENS by B mode, and allows the part of waveform of interest to be selected for expansion.

These are two of the methods used to produce the second, B sweep.

A display selection system is required on the front panel to complete the operation of this second timebase. There are several ways of showing the A and B sweeps either individually, or combined together, and the B sweep may be started either immediately after the delay time, or on the next trigger pulse after the delay time (see Chapter 8, Sects 8.2.2 and 8.2.3).

Now the display mode selector must provide at least three options in order to utilize the delayed timebase for the close examination of waveforms.

5.11   Display mode

First it must have a normal, or A ONLY mode, for standard (A) timebase operation. Then a combined mode, where the presence of the B sweep can be seen during the A sweep. The most popular way to achieve this is called A INTENSified by B. Here, during the A sweep across the screen, at the end of the delay time and for the duration of the B sweep, the trace is made brighter (intensified). So this highlighted portion is clearly visible on the screen. Now by adjusting the coarse (switch) and fine (potentiometer) delay time settings, or the Delay Time Multiplier, the brighter portion of the waveform can be set as the section scanned by the B sweep. This brighter 'B' section of the wave-form can then be expanded to fill the screen at the faster speed of the B timebase rate.

A third display mode is required to do this and is called 'B ONLY' or simply 'B'. So the three display modes are:

1   A (A only).
2   A INTENSified by B.
3   B (B only).

Other modes are often available, such as Mixed Sweep and Alternate Sweep.

5.12   Mixed sweep

This is a method of displaying the waveform at both the A (slower) and B (faster) sweep rates within one scan of the screen. The A sweep starts its scan from left to right as normal. Then (after the delay time) the B sweep starts and takes control of the horizontal deflection system. So the effect is to have a mixed display of slow scan on the left and fast on the right. The delay time multiplier is used as before to set the point on the A sweep from which the B sweep starts. The effect on the screen in the mixed sweep mode is to vary the point on the horizontal scan where the change occurs from slow (A) to fast (B) sweep rates. Figure 5.6 shows a square wave displayed in the mixed sweep mode. The A timebase speed is set to 2 ms/div, and the B sweep is set to 100 μs/div. The delay multiplier is set to 0.5, so the transition point from slow to fast speed is half-way across the screen.

5.13   Alternate sweep

Let us assume that Channel 1 only is displayed on the screen. When Alternate timebase display is selected, the Channel 1 trace is first displayed at the

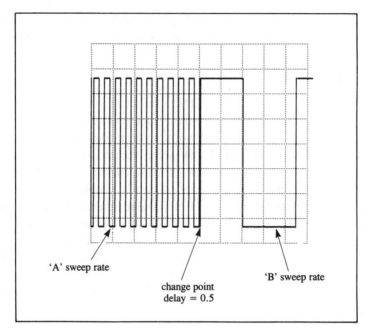

Figure 5.6. Screen display of a square wave signal in the mixed sweep mode.

normal A timebase rate, right across the screen. Then, on the next sweep, the same Channel 1 waveform is displayed right across the screen at the B time-base rate. Then the A sweep occurs, then again the B sweep, and so on. The A and B timebases alternately sweep the Channel 1 signal. So now we have two displays for the same Y channel: one at slow speed and one at high speed. To stop these two signals being confused due to them appearing at the same vertical position on the screen, a special circuit is used to introduce an offset voltage to the vertical system, so that on these alternate sweeps the trace is also alternately moved up and down on the screen. This is called the separation control and may be a front panel adjustment. So each time the A sweep occurs, the waveform may appear at the top of the screen, and on the B sweep the waveform may be at the bottom of the screen. So at high timebase speeds, two stable pictures appear, one an expanded section view of the other. On dual trace mode, there can be four displays: Channel 1 and Channel 2 at the A rate, and Channel 1 and Channel 2 at the B rate. So things can get rather confusing.

5.14   After delay trigger

As well as all the A and B sweep combinations above, a further refinement is the after delay trigger mode. Here a second 'B' trigger circuit is used to produce output pulses on each cycle of the displayed input waveform. When the delay time is completed, instead of the B sweep firing as above, it is only 'enabled' or made triggerable. So after the delay time, the next trigger pulse from the B trigger circuit will fire the B sweep. When high delay ratios are used, this give a much steadier display in the delayed mode, avoiding jitter (see Chapters 8 and 12 on after delay trigger).

5.15   Delay ratio

The delay ratio is the ratio of the delay time (from the start of the A sweep to the start of the B sweep) to the duration of the B sweep.

Some oscilloscopes have selectable delay time ranges, with coarse and fine controls and others have a delay time multiplier. In the latter case, a multi-turn potentiometer is used to set a ratio or percentage factor of the A sweep time (from the start point) to give the delay time. For example, if A timebase is set to 1 ms/div, the B timebase is set to 1 μs/div, and the delay multiplier set to 0.5 or 50 per cent, the B sweep will start half-way across the A sweep. The delay time will be 5 divisions (A) at 1 ms/div = 5 ms. So the delay ratio will be:

$$5 \times 10^{-3} : 1 \times 10^{-6} = 5000 : 1$$

High delay ratios may be considered to be in excess of 1000 : 1.

5.16   X amplifier

The X output or horizontal amplifier is much like the vertical output amplifier. It is a push pull, large signal amplifier, which drives the horizontal deflection plates of the CRT. There are two main differences between the horizontal and the vertical output amplifiers: the X amplifier has a larger voltage output swing since the CRT X plates are always less sensitive, and (partly as a consequence) the X amplifier has a lower bandwidth.

There are two inputs to the X amplifier; the d.c. voltage from the X position control, and the voltage ramp signal from the timebase. In the XY or horizontal external mode of the oscilloscope, the analog input is switched from the timebase ramp to the relevant X input source. On those oscilloscopes with X–Y mode, this may be the Channel 1 or Channel 2 vertical input amplifier, which now becomes the matched X input in this mode. On oscilloscopes with horizontal external input, the source may be from a separate external input socket, possibly via a pre-amplifier.

The bandwidth of the X amplifier does not need to be, and consequently is not as high as the vertical output amplifier, whereas the vertical amplifier system may have a bandwidth of, say, 20–100 MHz on the most popular models, the X amplifier bandwidth is typically 1–5 MHz. In normal operation this is obviously adequate for scanning the spot horizontally at the fastest sweep speed, but in the X–Y mode there will be a large difference between the

X and Y performance. Check the oscilloscope specification for the X band-width and X–Y phase response before using this mode.

5.17   X–Y mode

Most modern dual trace oscilloscopes now have the X–Y facility. When X–Y is selected, one of the vertical channels is used as the horizontal input, and the sweep generator is disabled. Now with both vertical input systems in use, say Channel 1 as the X and Channel 2 as the Y, there is a matched X–Y input system, with all the coupling, variable and attenuator controls available in both axes. Assume all Channel 1 and Channel 2 controls set to matched conditions, and a signal connected to Channel 2 which is a 5 V p.p., 1 kHz sinewave. With the attenuators set to 1 V/div, and both VARiables to CAL position, there will be a 5 division vertical line displayed in the X–Y mode (No X input). Now if the signal is transferred to Channel 1 (X) input, there is no vertical deflection, but a 5 division horizontal line displayed.

If the signal is simultaneously connected to Channel 1 and Channel 2 (X and Y) inputs, a diagonal line will be displayed, 7.07 divisions long. If you visualize this line as the diagonal formed from a right angle triangle with one side (vertical) 5 divisions high, and one side (horizontal) 5 divisions wide, then the length of the diagonal is the square root of (the sum of the squares of the other two sides):

$$= \sqrt{5^2 + 5^2} = \sqrt{25 + 25}$$
$$= \sqrt{50} \qquad = 7.07$$

There are many applications for the X–Y mode (see Chapter 8, Measurements) but there is an important limitation to remember. Although you may have a 20 MHz instrument (say) and the vertical channel you are using, here Channel 2, will thus be 20 MHz, the X channel will be much lower, perhaps 1 MHz. So limit the use of X–Y mode to signals below 1 MHz (at least for the X input). The same rules apply as for vertical amplifiers with regard to signals approaching bandwidth frequency (see Chapter 2, Sect. 2.12.3).

APPLICATION EXAMPLE

Measurement of 8 bit digital code

The object of this example is to set up and measure the binary code of an 8 bit digital word. Many computer and other digital systems operate with an 8 bit system. This is where each piece of information is encoded as a group of eight digital pulses, comprising a 'word'.

Each pulse or 'bit' can be at high (1) or low (0) level. In a positive logic system, TTL high is +5 V (1) and low is 0 V (0). In this example we shall look at the output from an 8 bit shift register and determine the binary code. In order to establish the timing, connect the Channel 1 probe to the clock pin of the shift register, and select Channel 1 trigger on the oscilloscope. Adjust the timebase and time VARiable controls to obtain exactly one cycle per division across the screen, but more significantly, the first eight clock pulses from left to right, to occupy exactly one division each. So the first 8 divisions from left to right are equal to the duration of one 8 bit word. This sequence displays on the oscilloscope screen the reference points where clock pulses occur so that we can compare each section of the unknown digital word and determine its high or low state. Leave the Channel 1 probe connected to the clock pin, but now select Channel 2 trigger on the oscilloscope. Connect the Channel 2 probe to the shift register output pin. Now when the 8 bit code appears at the output terminal synchronous with the clock on Channel 1, the data bits are displayed on Channel 2, as in the upper trace in Fig. 5.7.

Now since each division of the first eight (left to right) is equal to one bit time, it is simple to read the code from right to left. In this example the eighth bit is high (1), so it is easy to see when the eighth

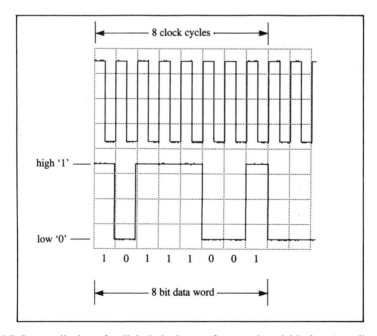

Figure 5.7. Screen display of a digital clock waveform and an 8 bit data 'word'.

pulse actually occurs. However, if the eighth pulse had been low, and possibly preceding bits also low, it would have been difficult to determine exactly when the word ended, without reference to the graticule lines and the clock reference above on Channel 1.

Looking at the first division on the left, the level is high	'1'
The second division is low,	0
The third division is high,	1
The fourth is high,	1
The fifth is high,	1
The sixth is low,	0
The seventh is low,	0
And the eighth is high,	1

So this data sequence, left to right, is   1 0 1 1 1 0 0 1

Taking the least significant digit to be on the right, we can convert this binary sequence to its decimal equivalent.

Bit no.	8	7	6	5	4	3	2	1
Decimal value	128	64	32	16	8	4	2	1
Code	1	0	1	1	1	0	0	1

So here where we have a '1' we can take the decimal values and add them.

128		32	16	8		1

So adding these values, $128 + 32 + 16 + 8 + 1 = 185$

The decimal equivalent of the measured binary coded sequence is thus 185.

6
Power supplies

Oscilloscope power supplies cover a wide voltage range from 5 V TTL level supplies, to 15 000 V or more for CRT acceleration voltages. Apart from the heater filament for the CRT, which is usually a 6.3 V a.c. supply, all power supply requirements are d.c. and usually stabilized. The use of stabilized supplies ensures the accuracy of the oscilloscope over different operating conditions, such as changes of sweep speed, display mode and brightness, etc. Also fluctuations in the mains input voltage (or battery voltage) and temperature will not affect the d.c. supplies. A stabilized supply is one whose output is controlled by taking a sample of the output voltage, and comparing it with a reference voltage. The control circuit feeds more or less power to the output circuit to maintain balance, so that the sampled output always equals the reference voltage. Thus the output voltage is maintained constant. Most oscilloscopes incorporate supplies which will cope with at least plus or minus 10 per cent input voltage variation around their nominal settings. Thus, if an instrument is connected for 240 V operation, it will maintain its calibrated accuracy at least from 216 to 264 V a.c. input.

There is a limited range of oscilloscopes available which operate from low voltage d.c. supplies. This may be from internal batteries fitted within the oscilloscope, or from an external d.c. supply such as a vehicle battery. These battery powered instruments have the advantage of total portability, and this enables them to be operated anywhere from the top of a mountain to the bottom of a mine. Liquid crystal displays (LCDs) are becoming increasingly popular in this field as they are both small and light, and also only require low voltage power supplies. Since their energy consumption is small, they are particularly suited to battery instruments as they allow long operating cycles between recharge or replacement of the batteries. Whether the power source for the oscilloscopes be from a.c. mains (line) supply, or d.c. batteries, the wide range of supply voltages inside the instrument means that a transformer is required in the system. A transformer can only operate with an a.c. input, so if the power source is d.c., such as a battery, it must be converted to a.c. first. This is done by using an oscillator powered by the d.c. supply. The oscillator output signal can then be used as the a.c. waveform for the primary input to the transformer. The a.c. input to the transformer can then be converted to the many different supplies required by the oscilloscope. This is achieved by transformer action, where the output or secondary voltages from the transformer are produced in ratio to the number of winding turns on the transformer secondary compared to the primary (input) winding. Broadly speaking, if the primary winding has (say) 1000 turns, and a particular secondary has 500 turns, then the secondary output voltage will be half the input voltage. These secondary a.c. voltages are then used to produce the d.c. voltages required for all the oscilloscope's circuits.

6.1   The d.c. supply voltages

The d.c. supply voltages fall into four categories which can only be broadly defined and will vary with each individual oscilloscope.

1 *Low voltage supplies* (5–30 V d.c.) A range of positive and negative supplies used to operate pre-amplifier circuits, trigger circuits, timebase logic and sweep generator circuits, etc. Often these low voltage supplies take the form of a monolithic regulator circuit, that is, a complete stabilized voltage supply in one package or 'chip'. These regulators may sometimes have only three pins, and look like a single transistor, but in fact contain a complete set of control and output circuitry, housed in one unit.

2 *High voltage supplies* (100–300 V d.c.) There are normally two main supplies in this range, one for the X deflection and one for the Y deflection circuits for the CRT. These voltages are also used for the correction of CRT geometry, astigmatism, and so on. These supplies are usually in the form of discrete circuitry, that is separate individual components, laid out on a printed circuit board (PCB). Because of the higher voltage and power requirements, these supplies are not commercially available as monolithic regulators.

3 *Extra high tension (EHT) supply* (−500 to −3 kV) The cathode of the CRT is operated at a negative potential between about −500 V and −3000 V d.c., depending on the type of CRT used. Associated with the EHT supply, some oscilloscopes employ a low voltage d.c. supply of about 30–50 V (with respect to cathode *not* to ground). This is used for the blanking or bright up circuitry and effectively gives a second EHT supply 20–50 V above (less negative than) the main EHT voltage.

   The difficulties of controlling power supplies with outputs of several thousand volts, means that complex circuitry must often be used. Special circuit techniques are employed to control very high voltages using relatively low voltage devices. However, the principle of operation is the same; a sample of the output voltage is compared with a fixed reference voltage, and the controller supplies more or less power to the output to maintain balance at the comparator.

4 *Post-deflection anode (PDA)* (3–25 kV) Oscilloscopes with PDA CRTs require a further very high voltage positive supply. This is connected to the final anode located at the front of the CRT and provides the extra accelerating potential required for a very bright trace with high writing speed. The level of voltages involved makes these circuits very hazardous, and they should be carefully avoided by untrained personnel. The high voltage is most often developed by using a voltage multiplier system of capacitors and diodes to reach a d.c. potential of (say) 15 kV (15 000 V) from the transformer secondary output of (say) 1000 V. This might be the same transformer connection as used for the EHT supply, as above.

6.2   Mains (line voltage) supplies

Traditional power supplies, which are probably still the most common, utilize a mains transformer (see Fig. 6.1). This is used to provide many output (secondary) windings, each with an a.c. output voltage to suit each part of the circuit. These a.c. secondary windings are developed from the turns ratio of the transformer multiplying or dividing the mains (primary) voltage input. Each secondary output voltage is then full- or half-wave rectified, and fed to a feedback control circuit to provide a stable d.c. output voltage supply. Half-wave rectified means that the a.c. waveform from the transformer is supplying power during only half of the cycle. A rectifier is used to turn this a.c. half-cycle into a d.c. voltage (together with a capacitor). The rectifier diode might be connected so that it conducts on every positive half-cycle, and is reverse biased, or 'cut off' during the negative half-cycles. Alternatively, the negative half-cycle might be used for the supply, while the positive half-cycle is unused. In the case of full-wave rectification, both half-cycles are used, each turning

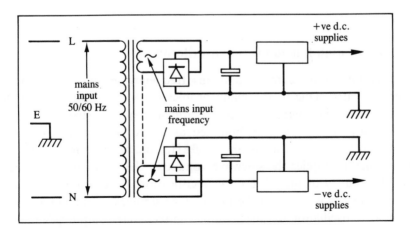

Figure 6.1. Simplified diagram of a typical mains transformer, regulated, d.c. power supply.

on a separate diode. In fact more diodes and more capacitors are used, but the principle is the same.

This stabilized supply will then be constant irrespective of mains input voltage variation, load current or temperature variations. Since this system operates at mains frequency, which is 50 Hz in the United Kingdom and 60 Hz in the United States, the secondary voltages developed by the transformer are also at 50 Hz (or 60 Hz). Thus when these supplies are rectified and smoothed to provide d.c. supplies, large capacitors are required as reservoir and smoothing components. The electrical capacitive (and physical) size of the capacitors used is inversely proportional to the operating frequency. The higher the frequency, the smaller the capacitor. At the relatively low frequency of 50 or 60 Hz, the capacitors need to be large to achieve the required reduction in ripple levels. Typical values are 50–1000 μF and consequently they are large and heavy. At the higher frequency of 50 kHz (1000 times higher) for the same smoothing factor, the capacitor can be 1000 times smaller – 50 nF to 1 μF for the same ripple voltage. The ripple voltage is the amount of a.c. voltage remaining on the d.c. supply. The a.c. voltage system must be used in the first place so that transformer action can produce all the different voltage outputs required, but once the supply has been rectified to d.c., the a.c. voltage content must be made as small as possible. The objective is to have d.c. only, with no a.c. (ripple) superimposed on it. In practice this is impossible, so realistic levels of ripple that are designed for are less than 0.1 per cent. That is, a ratio of 1000:1 d.c. to a.c. (ripple). So for a d.c. supply line of (say) 10 V, the ripple level, or a.c. present on the 10 V rail, should be $(0.1/100) \times 10 = 10$ mV p.p. maximum.

Similarly, for a 100 V d.c. line, the ripple voltage should be less than 100 mV p.p. These are typical figures only, and some circuits may demand lower (or higher) ripple levels.

6.3   Extra high tension (EHT) oscillator

One exception to these rectified mains frequency secondaries is the high voltage oscillator. Due to the many problems associated with high voltage supplies, one method used is to operate a high frequency sinewave oscillator from the low voltage supplies obtained as above. This oscillator operates above audio frequency, usually about 30 kHz, and feeds a small EHT transformer which multiplies the sinewave voltage to the required level to supply the EHT circuitry of the CRT (after rectification). The primary side of the EHT transformer is fed from the low voltage supplies (as above) and results in a sinewave oscillation of 20–60 V p.p. The secondary output side magnifies the sinewave voltage to 500–1000 V p.p. At this high frequency, it is much easier and more compact to smooth the rectified voltage to the required ripple

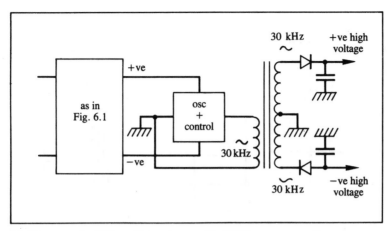

Figure 6.2. The use of a HF oscillator to provide EHT power supplies.

level. At this frequency the smaller capacitors used give great space and cost benefits as well as the greater safety factor since the stored energy is much less. The long-term reliability is also improved due to the use of non-electrolytic capacitors (see Fig. 6.2).

6.4 Switched mode power supplies (s.m.p.s.)

The switched mode power supply is becoming increasingly popular due to the demand for lighter more compact oscilloscopes with more performance, particularly digital storage scopes.

The switched mode power supply operates at a frequency just above the audio band, usually about 30 kHz (see Fig. 6.3). The switched mode system operates with a pulse type waveform rather than a sinewave, and by controlling the pulse width as the output loading or input supply voltage varies, the supply can be maintained constant. When the battery or mains input supply to the oscilloscope is at its nominal value, the s.m.p.s. waveform is virtually a square wave. If the input supply level falls, or the load demands more power, then the width of the pulse supplying energy increases. For instance, in the case of a positive supply rail, the positive going pulse in the switching waveform would increase its width until the d.c. rail was restored to its nominal value. So as the input changes from a very low to a very high limit, the positive pulse (in this case) changes from very wide to a very narrow pulse width. Figure 6.4 shows typical primary oscillation waveforms from a switched mode power supply. The upper trace is the normal waveform when the mains supply is at its nominal 240 V level. The lower trace shows the same switching waveform when the mains supply has fallen to about 150 V.

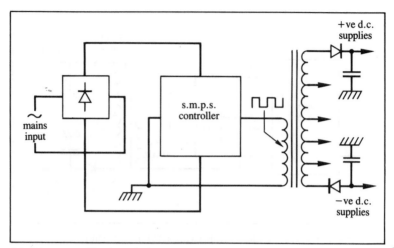

Figure 6.3. Diagram of basic switched mode power supply system.

The mains supply is first rectified to a d.c. voltage, and this d.c. supply operates the 30 kHz oscillator. The output from this oscillator is fed to a transformer primary, and the secondary output winding provides all the different output voltages required by the oscilloscope.

Now there are many advantages to this type of power supply. Since the system operates at a high frequency, the transformer size can be made very small with high efficiency and great weight reduction and less heat dissipation. By regulating the switching cycle of the supply with a control system, all the secondary voltages can be stabilized at once, and this one controller can maintain the output constant for very wide variations of input voltage. Often these power supplies can handle input voltage variations of more than plus or minus 50 per cent. So if the mains selector was set to 200 V, the oscilloscope could be operated with mains input voltages from 100 to 300 V, and on the 100 V position, from 50 to 150 V. So an enormous mains supply variation can be accommodated with just two selector settings. At the high oscillator frequency, reservoir and smoothing capacitors for the d.c. supplies on the output side can be much smaller for the same ripple levels, so overall, the switched mode supply gives enormous savings in weight, size and efficiency.

One major disadvantage is the fact that the switching waveform in the supply exhibits sharp 'spiky' pulses, and these tend to radiate everywhere, so care must be taken to avoid interference from these spikes both inside and outside the oscilloscope, and also back into the mains supply (Fig. 6.3). Special filters must be used to prevent any form of interference from passing back into the mains supply, and these interference levels are becoming increasingly controlled by international agreements and legislation.

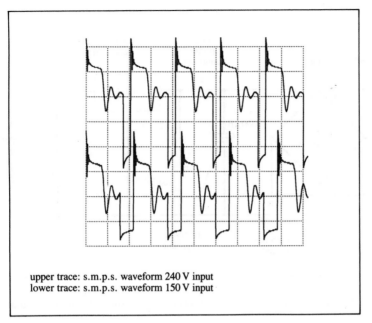

upper trace: s.m.p.s. waveform 240 V input
lower trace: s.m.p.s. waveform 150 V input

Figure 6.4. Typical s.m.p.s. output waveforms, showing the effect of variation of mains input voltage.

APPLICATION EXAMPLE

Diagnosis of faulty power supply – faulty rectifier

In this example we shall look at a faulty power supply circuit to determine the fault. Figure 6.5 shows the circuit diagram of a +5 V stabilized power supply using a 3 pin 7805 type regulator. The secondary winding of a transformer feeds a full wave bridge rectifier, which supplies a 1000 μF reservoir capacitor, and provides about +9 V d.c. to the input of the regulator. The +5 V d.c. output from the regulator should have less than 10 mV p.p. ripple voltage, but under load conditions the input terminal should have a ripple level of about 2 V p.p.

Additional 10 nF low loss capacitors are fitted to eliminate high frequency noise and oscillation on the supply rail. The +5 V output rail is known to be faulty, as a 50 Hz waveform is appearing on the 5 V line supplying the TTL integrated circuits in the load circuit.

First connect the oscilloscope probe to the output terminal of the regulator. Figure 6.6 shows the pulse at the regulator on the lower trace (Channel 2). The Channel 1 probe is then connected to the input terminal of the regulator, and the upper trace of Fig. 6.6 shows the

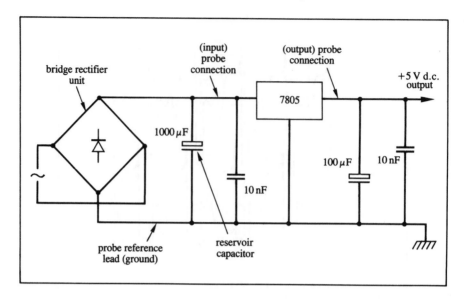

Figure 6.5. Circuit diagram of a 5 V d.c. power supply using a three pin regulator.

input waveform. Although Channel 2 was a.c. coupled just to observe the waveshape, set the Channel 1 input coupling switch first to ground and use the Channel 1 position control to set the trace line to a lower graticule line as the reference point (0 V). Then set the Channel 1 coupling switch to d.c. The upper trace in Fig. 6.6 shows the waveform. There is a large 50 Hz signal varying between +3 and +7 V. Since the minimum level (3 V) of this voltage is below the minimum input level for the regulator, the regulator cannot maintain the output up at +5 V, so the ripple transfers to the output.

There are many possible causes for this large ripple voltage across the reservoir capacitor, but a clue is in the frequency. Set the timebase switch to 5 ms/div and measure the distance between two of the peaks of the waveform. The period is measured as 4.0 divisions, and with a timebase speed of 5 ms/div, this gives the period of the signal as $4.0 \times 5\,\text{ms} = 20\,\text{ms}$. Now the frequency is the reciprocal (1 divided by) of the period, so the frequency works out as

$$\frac{1}{20 \times 10^{-3}} = 50\,\text{Hz}$$

The ripple voltage from a full wave 50 Hz supply should be 100 Hz. In this case the ripple voltage is at 50 Hz.

In a full wave system, the bridge rectifier conducts on both half-cycles of the secondary waveform from the transformer, so the 50 Hz

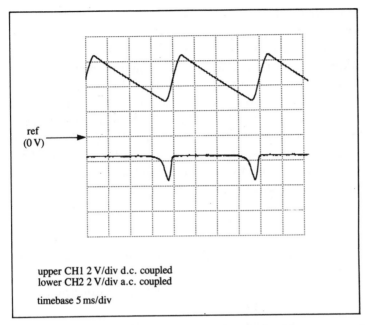

upper CH1 2 V/div d.c. coupled
lower CH2 2 V/div a.c. coupled

timebase 5 ms/div

Figure 6.6. Screen waveforms of the ripple voltages in a faulty d.c. power supply.

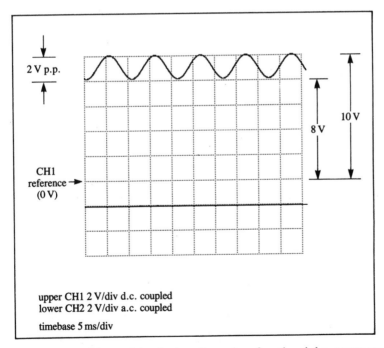

upper CH1 2 V/div d.c. coupled
lower CH2 2 V/div a.c. coupled

timebase 5 ms/div

Figure 6.7. Screen waveforms of the ripple voltages in a functional d.c. power supply after repair.

input to the bridge produces an output on each *half*-cycle. So a 100 Hz ripple is produced.

Here, in the faulty circuit, the ripple at 50 Hz thus indicates that the bridge rectifier unit is faulty, and should be replaced. After replacement of the bridge rectifier, the Channel 1 probe is reconnected across the reservoir as before, d.c. coupled, after setting the ground reference position on the graticule (see Fig. 6.7). Now the peak-to-peak value of the ripple voltage has fallen to about 2 V p.p. The d.c. value of the voltage has increased to about +9 V, and the period of the waveform has reduced to two divisions. This gives a period of $2 \times 5\,\text{ms} = 10\,\text{ms}$, and hence a frequency of

$$\frac{1}{10 \times 10^{-3}} = 100\,\text{Hz}$$

Having cured the input voltage fault, now reconnect the Channel 2 probe to the regulator output, a.c. coupled. Now the output ripple level has fallen to a low voltage which is not discernible, as shown on the lower trace in Fig. 6.7.

So now the circuit is functioning correctly.

7
Cathode-ray tube (CRT)

The heart of the oscilloscope is the display screen itself – the CRT. The CRT is a glass bulb which has had the air removed and then been sealed with a vacuum inside. At the front is a flat glass screen which is coated inside with a phosphor material. This phosphor will glow when struck by the fast moving electrons and produce light, emitted from the front and forming the spot and hence the trace. The rear of the CRT contains the electron 'gun' assembly. A small heater element is contained within a cylinder of metal called the cathode. When the heater is activated by applying a voltage across it, the cathode temperature rises and it then emits a stream of electrons. Now the cathode is connected to a high negative potential, and the deflection plates are connected to a positive voltage (see Fig. 7.1). The figure shows the layout of a typical p.d.a. cathode-ray tube. Although a post-deflection anode (PDA) type of CRT is shown, with a very high positive final anode voltage, the voltages and construction are much the same for a monoaccelerator tube except, of course, that there is no PDA electrode. Monoaccelerator means that there is only one (mono) acceleration voltage system in the tube, that is the voltage between the cathode and the group of electrodes comprising A1, A2, A3 anodes and the deflection plates.

The cathode is typically connected to $-2000\,V$, and deflection plates normally are at about $+100\,V$. So the electrons emitted from the cathode are attracted towards the positive voltage on the deflection plates, and so travel at high speed between and beyond these deflection plates to the front of the CRT. Between the cathode and the deflection plates are many correction electrodes, such as the focus and geometry electrodes which allow the user to adjust for a bright sharply focused even picture on the screen. The only elements which need concern the beginner are the cathode to emit a stream of electrons, the X and Y deflection plates to move the beam sideways and vertically, and the phosphor coated screen to make the electron beam visible.

Some (higher performance) oscilloscopes also have a post-deflection accelerator (PDA) at the front of the tube just behind the faceplate. When a very high positive voltage of say $10\,kV$ ($10\,000\,V$) is connected to this anode, the electron stream is enormously accelerated towards the phosphor screen. This results in a much brighter spot or trace on the display, and enables the spot to

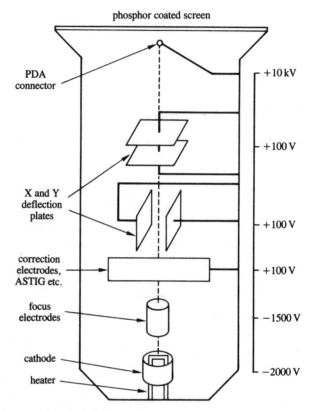

Figure 7.1. Diagram of typical CRT construction.

be scanned across the faceplate at much higher speeds while still remaining brightly visible.

The way that the electron beam is controlled in the tube is very similar to that of a light beam using optical lenses. The source of the beam is the electron gun at the rear, and the correction electrodes are like the lenses focusing and shaping the beam to produce a fine point of light on the screen. We shall consider each section of the tube in a little more detail to see just how it works.

7.1   Electron gun

The beam of electrons is produced at the rear of the tube in the 'gun' assembly. This comprises a triode or three-electrode system, just like a triode thermionic valve in a pre-transistor radio. The three electrodes are: the cathode, which actually emits the electron stream; the grid, to control the flow of these electrons; and the first anode, to attract and accelerate the beam towards it. Like all valves, the system operates within a vacuum, that is, all the air has been removed from the tube prior to sealing. If air were present, the action of the heater would cause combustion, and the tube would destroy itself. When the cathode and grid are at the same potential, the electron stream released from the cathode is attracted to the positive voltage on the first anode. If the grid voltage is made slightly negative with respect to the cathode, some of the electrons are repelled back towards the cathode, so the amount of electron beam current passing through the grid on to the anode is reduced. If the grid is made sufficiently negative with respect to cathode, all the electrons are repelled back to the cathode and hence the beam is cut off. In this condition, there is no beam current flowing down the tube towards the first anode.

7.2   Control grid

The control grid, situated between the cathode and first anode, is used to adjust the amount of electrons reaching the anode. The electrons emitted from the cathode are accelerated from the negative potential on the cathode, to the positive potential on the anode. However, if the grid is made more negative than the cathode, then the electrons are repelled back towards the cathode. If the grid voltage is made sufficiently negative to cathode, the electron stream is cut off completely. Figure 7.2 shows the effect of varying the grid voltage from zero with respect to cathode, to 50 V negative to cathode. The beam current reduces from 5 mA to zero. The front panel intensity control is used to vary the grid to cathode voltage in this way, hence adjusting the trace brightness from maximum to minimum.

The control grid is also used to turn the beam on and off during the sweep

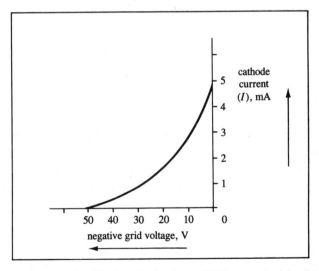

Figure 7.2. Graph showing the relationship between CRT control grid voltage, and cathode current.

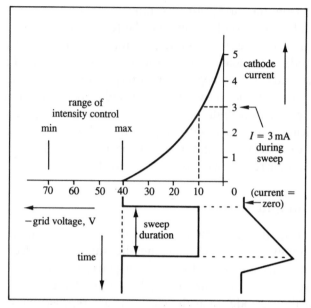

Figure 7.3. The relationship between the unblanking pulse, intensity control and grid voltage/cathode current characteristic.

cycle, making the trace visible during the sweep. In the flyback, hold-off and waiting times, the beam is turned off so that the trace is not visible. The unblanking pulse applied to the grid is positive during the sweep time, and negative for the rest of the cycle. Since the intensity control sets the grid negative to cathode, the positive unblanking pulse makes the grid *less negative to the cathode*. But it does not make the grid positive with respect to cathode (see Fig. 7.3).

The control grid is always operated negative with respect to cathode, so the combination of the bias potential set by the intensity control and the excursion of the unblanking pulse always remain in the region negative to cathode.

When the intensity control is operated in practice, it is during the sweep cycle, when the trace is visible. It sets the cathode to grid negative bias level from which the unblanking pulse rises, to make the trace visible. If the timebase were to stop working, the bias set by the intensity control would keep the trace blanked out, even if set to maximum. Hence the name of the unblanking pulse applied during the sweep portion of the timebase cycle to render the trace visible.

7.3   Focus

When the grid is at or close to the cathode potential, the electron stream leaving the cathode is attracted to the high positive voltage on the first anode, A1.

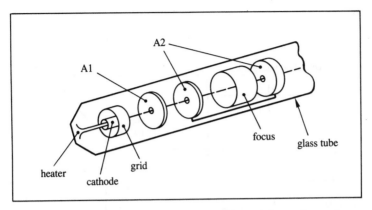

Figure 7.4. The construction of the CRT gun assembly and focus electrodes.

The beam of electrons is fairly broad and not parallel, so by the time it reached the faceplate and struck the phosphor, it would give a large spot of light on the screen. So a cylindrical electrode is placed in the path of the beam, like a metal pipe through which the beam passes. This focus electrode has a high negative potential connected to it, and has the action of condensing the electron stream into a fine narrow beam. The beam passes through the centre of the focus cylinder, and the repulsion effect of the electrostatic field within it concentrates and hence focuses the beam. By careful adjustment of the focus control on the front panel, the spot of light can thus be made sharp, round and bright. The focus electrode is actually situated between the two halves of the second anode, A2. It may take the form of a solid metal cylinder, or a series of metal discs in line, electrically connected together. The second anode, A2, is at about the same potential as A1, so after the electrons are attracted to A1, they pass on through A2. The middle of A2 contains the focus electrode, with its high negative voltage concentrating the beam. Then the forward half of A2 accelerates the beam onward towards the next elec-trodes (see Fig. 7.4).

7.4   Deflection plates

The next group of control electrodes that the electron stream encounters as it passes through the tube is the deflection system. The first to be met are the more sensitive vertical deflection plates, then, nearer to the screen and hence less sensitive, the horizontal deflection plates.

The push pull analog voltage waveform that is applied across the vertical deflection plates should be an exact (amplified) version of the signal connec-ted to the Y input socket. The amount of voltage applied to the deflection

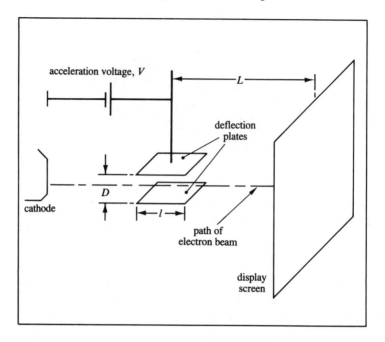

Figure 7.5. Simplified diagram of the mechanical relationship of the deflection system in a CRT.

plates for each centimetre of vertical screen deflection is dependent upon several factors (see Fig. 7.5).

● The longer the length of the deflection plates (l), the less voltage required.
● The smaller the distance (D) between the plates, the less voltage required.
● The greater the distance (L) from plates to screen, the less voltage required.
● The smaller the voltage (V) between the cathode and the plates, the less voltage required.

These factors are linked by the formula

$$\textbf{Deflection sensitivity } (S) = \frac{l \times L}{2D \times V} \text{ cm/V}$$

The reciprocal gives the **deflection factor** (DF) in volts/cm

$$DF = \frac{2 \times D \times V}{l \times L} \text{ V/cm}$$

So, for example, if the distance between plates $= 0.2$ cm, the cathode to plate voltage $= 2000$ V, the length of the plates $= 3.5$ cm and the plate-to-screen distance $= 20$ cm, then the deflection factor

$$DF = \frac{2 \times 0.2 \times 2000}{3.5 \times 20} = \frac{800}{70}$$
$$= 11.4\,\text{V/cm}$$

So with a vertical scale of (say) 8 cm, and allowing another 1 cm above and below the graticule, to get to the edge of the screen, 10 cm of deflection is required. So the vertical deflection voltage required would be $10\,(\text{cm}) \times 20\,(\text{V/cm}) = 200$ volts.

Of course the tube is designed to optimize the performance of each section. But, as always, the design is a compromise of opposing factors. For instance, to make the Y plates as sensitive as possible, to minimize deflection voltages, the plates should be large. But increasing the plate size increases their capacitance, and hence limits the high frequency performance. Thus for capacitance considerations, the plates should be small. So the plate size is a compromise between maximum sensitivity and minimum capacitance. In practice, special design techniques are used such as curved and tapered deflection plates to optimize their performance.

After the vertical deflection system, the beam next encounters the horizontal deflection system, which works in exactly the same way as the vertical. Obviously the horizontal plates are mounted at 90 degrees to the vertical plates, being each side of the CRT.

The X plates do not have to scan as fast as the Y plates, and therefore can be less sensitive. Being nearer to the CRT screen, but usually longer, the X plate deflection factor might be (using the same formula as before) as follows: Distance between plates $= 0.5$ cm, voltage $= 2000$ V, plate length $= 5$ cm, plate to screen length $= 16$ cm.

$$DF = \frac{2 \times 0.5 \times 2000}{5 \times 16} = \frac{2000}{80} = 25\,\text{V/cm}$$

So you can see that although the X plates are bigger than the Y, the shorter distance to the screen means that twice as much voltage is required on the X plates to move the spot 1 cm on the screen.

After passing between the X plates, the beam then continues forward until it strikes the front screen. In high performance tubes, there is usually another electrode between the X plates and the front screen, which is the dome mesh. This is an electronic lens, dome shaped, and constructed of fine wire mesh. As the beam passes through this lens, the angle of the beam is magnified from the centre line, so that the deflection is increased. This has the effect of magnifying the sensitivity of the deflection plates, so that the beam is deflected further on the screen for the same voltage applied to the plates. Although it has the great advantage of increasing the deflection plate sensitivity, it unfortunately interrupts a large proportion of the beam current, hence preventing it from reaching the screen. Consequently, these scan magnification systems are found in association with PDA tubes in order to maintain a high trace brightness.

7.5   Post-deflection anode (PDA)

The post-deflection anode (or accelerator) is the final electrode mounted at the front of the tube, giving the last high acceleration to the beam. This anode is usually connected to a potential in excess of 10 000 volts, and provides the beam with enough velocity to produce a very bright, yet sharp trace on the screen. These PDA-type tubes are usually found in high performance oscilloscopes where high bandwidth and fast timebase speeds are used. Also they are almost essential in delay timebase instruments in order to produce sufficient brightness with high delay time ratios.

7.6   Graticule

The graticule is the ruled grid which appears in front of the trace and facilitates measurements. The most common is the 10×8 cm graticule, giving 10 cm of horizontal and 8 cm of vertical measurement capability. Traditionally, the CRT was made separately from the graticule, and in fact the graticule usually took the form of a small plastic sheet, marked with the graticule lines. The plastic graticule was then mounted in front of the CRT faceplate, with the marked side (the face of the graticule actually bearing the lines) back towards the glass faceplate. This was to minimize the parallax error when making measurements. Imagine a stationary dot in the centre of the screen, and you view this dot with one eye directly in front of the dot and the other eye closed. Now keeping your head still, open the other eye. Now with this eye you will see that the spot is no longer aligned with the screen centre line. This is because, due to the thickness of the CRT faceplate itself, about 5 mm, there is a gap between the actual spot on the phosphor inside the CRT, and the graticule line. Now in practice, with a full 10 cm trace across the screen, the measurement you take will depend upon the position of your eyes relative to the graticule, and whether you move your head while taking the measurements. The same problem obviously occurs in the vertical direction. Now this problem has been overcome in many oscilloscopes by using a CRT with an internal graticule. Here, the graticule lines are marked on the inside of the faceplate under the phosphor so that the trace is actually formed in the same plane as the graticule. The graticule lines are marked on the glass faceplate first, then the phosphor coating is added, prior to the construction of the CRT. Now with no gap between the trace and the graticule, no parallax errors are possible.

7.7   Filter

A blue-green filter is fitted directly in front of the CRT faceplate to optimize the contrast ratio of the trace on the screen, that is, to make the trace as bright as possible and the background as dark as possible. The colour of the filter is chosen to match the light output from the particular CRT used, and for normal persistence CRTs this is the green to blue colour. For long persistence CRTs, an orange filter is used to match the spectral response from the phosphor compound used to give the long afterglow.

The persistence of the CRT is the time it takes for the light output to fall to a certain level, usually 10 per cent of the original brightness, when the beam is cut off, and the most commonly used are the P1 or P31 type phosphors. These have a decay or persistence time of less than a second in practice, with normal ambient lighting. The longer persistence P7 type used with the orange filters has a much longer afterglow, but again in normal ambient lighting conditions, it is two or three seconds. There are even longer persistence types available for special applications, but these are becoming less common now due to the introduction of the low cost digital storage oscilloscope.

7.8   Screen readout

A useful feature becoming widely available in lower cost oscilloscopes is the screen readout display. Alphanumeric characters are displayed at the top and bottom edges of the screen to indicate various data. The data presented show the status of the instrument, and sometimes also measurement data from the displayed waveform. Waveform measurement data are mainly found in conjunction with cursor measurement, where electronically generated cursor lines can be positioned on the screen.

The status data most often presented are timebase and attenuator settings, trigger levels, etc. These parameters are particularly useful when photographic recordings are made of the screen, as the photos then show all the parameters as well as the waveforms. Also when the oscilloscope waveforms are transferred after storage in digital format (in digital storage oscilloscopes) to other equipment such as computers and printers, then these front panel parameters can be transferred with the waveform data. This enables a printer to record the settings at which waveforms were displayed, and computers can process the waveform data they receive, having the scaling factors for the waveforms. Cursor measurements are made by positioning of pairs of horizontal and vertical lines on the displayed waveform. If one vertical cursor line is positioned at the start of one cycle of the displayed waveform, and the second vertical cursor positioned at the end of the cycle, the time between the two lines is the period of the waveform. This time period is calculated within the oscilloscope and then displayed in the data readout area. Some models also allow further calculations from these data, such as frequency readout (the reciprocal of the period) at the touch of a button.

A similar procedure is used in the vertical direction, positioning a horizontal cursor line at the top and another at the bottom of the displayed waveform. The distance between is the peak-to-peak amplitude of the waveform, and this is calculated in the oscilloscope and displayed in the readout area.

Thus each pair of cursor lines is used as two reference points for waveform measurement. Since all the measurements are made on the display, without need to check front panel settings, these cursor measurement systems are ideal for inexperienced or occasional oscilloscope users. Provided that all the front panel VARiable controls are set to their CALibrated positions, time, frequency and voltage amplitude values can be read directly from the screen. Uncalibrated setting of the variable controls may be indicated on the screen readout display, but even so, care must be taken to check for this condition, and if necessary remedy it. Although the screen readout characters may have a dot format appearance, they are formed on the screen by the scanning of the beam, in the same way as the waveform display, so ordinary CRTs are used.

7.9   LCD displays

A system coming more into use in the modern oscilloscope is the liquid crystal display (LCD). Instead of using the vacuum CRT, with its large size and weight and associated high voltages, a low voltage, flat screen system can be used. So why not use these displays in all oscilloscopes? There are three main reasons at the moment, which may with technological development, soon disappear.

1 They are expensive compared to a CRT.
2 The resolution is not as good. Although the actual dot or pixel size is quite small, being about 0.3 mm at present, the problem is the gaps between the dots, which give the display the 'broken up' or dotted appearance. With the CRT, the phosphor coating is continuous, so the display is a much more acceptable unbroken presentation.
3 The third reason is the writing speed of the display. Because of its digital scanning requirement, to operate the LCD, the analog signals have to be converted to digital format. This obviously lends itself to digital storage, but gives the problem of not seeing the signal in real time. Also the bandwidth of the LCD system can only be as fast as the digital a to d (analog to digital) and d to a converters allow.

For the moment, CRTs continue to be the usual oscilloscope waveform display system.

7.10   Trace rotation

When a line (no vertical input) is displayed on the screen of the CRT, it is often found that it is not exactly parallel with the horizontal graticule line. This is due to the earth's magnetic field influencing the accelerating beam inside the CRT. The scanning system in a television works by using coils to produce a changing magnetic field to scan the beam in a TV tube. However, in the oscilloscope, the magnetic field effect from the earth itself, although very small, is a nuisance. To overcome this effect, a coil is fitted around the CRT, and an adjustable d.c. current passed through it. By setting the front (or rear) panel control (Trace Rotate) the coil current can be adjusted to cancel out the earth's effect. This is done by simply aligning the trace with the graticule line. It may be necessary to readjust the control if the instrument is moved and its orientation changed.

APPLICATION EXAMPLE

Measurement of monostable time constant

The object of this example is to measure the time period or time constant of a monostable multivibrator. A monostable is a form of oscillator circuit with only one (mono) stable state. It usually has two outputs, in antiphase, Q and $\overline{Q}$. The stable state, when the mono-stable is untriggered, is with Q at '1' level (high) and $\overline{Q}$ at '0' level (low). The monostable in this case is triggered by a positive going edge on its input terminal. When the monostable is triggered, its outputs change states, Q going low and $\overline{Q}$ going high for the duration of the monostable time constant. This time constant is set by the addition of external timing capacitor Cx and resistor Rx connected to the appropriate pins of the device. First the Channel 1 probe is connected to the trigger input terminal, and the oscilloscope set to Channel 1 trigger. The Channel 2 probe is connected to the $\overline{Q}$ output terminal. The oscilloscope trigger is set to positive (+) polarity, so that when the input trigger pulse into the monostable rises, the time-base sweep starts.

   Figure 7.6 shows the trigger pulse input edge on the upper trace. The lower trace shows the resultant timed output at $\overline{Q}$. The timebase switch is set to the fastest speed possible where the *whole* of the posi-tive pulse at $\overline{Q}$ is displayed on the screen. The time VARiable is set to the CALibrated position. The X position control is then adjusted to align exactly the leading edge of the output pulse with the first grati-cule line on the left. The pulse width is then measured against the

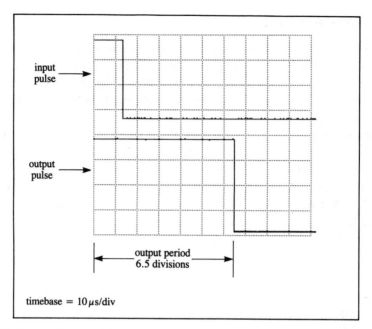

input
pulse

output
pulse

output period
6.5 divisions

timebase = 10 μs/div

Figure 7.6. Screen display of the input and output pulses in a monostable circuit.

graticule to the falling negative edge of the pulse. In Fig. 7.6 this width is 6.5 divisions. The timebase range was set to 10 μs/div, so the pulse width is thus

$$6.5 \times 10 \, \mu s = 65 \, \mu s$$

So the time constant for this monostable is measured as 65 μs (65 microseconds).

8
Measurements

There are two uses for the oscilloscope CRT screen: as a display device and as a measurement system. Much information can be gained from simply looking at the waveforms displayed on the screen: shape, size, stability, etc. Also comparing two waveforms (or more) on dual or multitrace instruments indicates relative timing and even presence of information in the waveforms. This can all be done by observation alone, where the operator, using experience and logic, can evaluate the information presented.

For many applications, however, it is necessary to make careful and accurate measurements of various parameters of the waveform, and in this case it is vital to know how to make and interpret these measurements.

The two general categories of measurements to look at are:

1 Vertical measurements
2 Horizontal measurements

We can divide the first category into several areas which we shall look at in turn. Before making any measurements, make sure that any variable controls are in the calibrated (CAL) position.

Before actually making any measurements, it is important to remember the effects of the probe (you should be using a probe) loading on the circuit you are testing. For most measurements, the probe will have no noticeable effect on the circuit conditions, but when looking at high frequencies or fast risetimes, the probe may cause considerable changes and give false results. As stated in Chapter 3, always use a ×10 probe where possible. This will have a load of about 10 Mohm and 12 pF on the circuit under test. As well as possibly changing the circuit conditions when the probe is connected, when making risetime measurements, the bandwidth of the probe itself may have a large effect on the measurement (see Sect. 8.2.1, Risetime measurements).

On some (usually more expensive) oscilloscopes the probe attenuation factor is automatically adjusted by the oscilloscope provided that the manufacturer's recommended probes are used. This may be in the way that the sensitivity is indicated on the front panel, the screen readout, the cursor measurement system and/or the data output to peripheral devices. However, on the general purpose oscilloscope, it will be necessary to include the probe attenuation factor when measuring voltages.

Although, in general, the only two measurable parameters on an oscilloscope are voltage and time, there are many other dynamic effects that can be displayed and measured on the oscilloscope with the help of an appropriate transducer. This is a device or system that will convert the phenomenon being monitored into a voltage. These could be such things as pressure, sound, light, heat, magnetism, or virtually anything that can be converted into a voltage (and most things can). In order that the transducer gives meaningful results, it must have a calibrated correlation between its input parameter and its output voltage. A very simple example is the current probe. Here the input electrical current is related to the output voltage as milliamps per millivolt (mA/mV). For example, if such a probe has an output of 5 mA/mV, then if connected to the oscilloscope on the 1 mV/div attenuator range, 5 mA p.p. of current will produce 1 div p.p. on the screen. If the attenuator was set to the 5 mV/div range, then 25 mA would give 1 div display (or 5 mA current would give 0.2 div display). There is a fuller explanation in Chapter 3. Except when the oscilloscope is used in the X–Y mode, whatever the original change, heat, light, current, etc., it is always measured against time, so the following horizontal measuring techniques are always the same. In the case of the vertical input, care must be taken to transpose the parameter ratios accurately.

When the X–Y mode is used, the same care must be taken with the horizontal input scaling, since the timebase is not in use and the horizontal deflection will usually represent some parameter other than time.

8.1   Vertical measurements

8.1.1   Alternating current and direct current

Although not always possible, the most comprehensive way to obtain information about a waveform is to feed it into the oscilloscope vertical amplifier d.c. coupled. Now as well as being able to measure the size of the signal, we can also see and measure any d.c. content. Before connecting the signal, the trace should be set exactly on a horizontal graticule line as a reference point. The centre line is often the best place to start. Set the variable controls on the vertical amplifiers being used to their calibrated (CAL) positions and set the attenuators to the least sensitive position (fully counterclockwise). Now connect the signal to the Y input and watch what happens. It may now be necessary to turn the attenuator clockwise to increase the display size to a practical level. This should be as large as possible within the screen area. If the waveform moves off the top or bottom of the screen before the signal display becomes sufficiently large, this indicates the presence of d.c. voltage content. If the waveform is equally disposed about the centre line, there is no d.c.

content. (This is not the case when the input coupling switch is in the a.c. position.)

Firstly, let us suppose that there is a d.c. voltage present on the signal. When you first connect the signal, watch which way the trace moves on the screen. If it goes upwards (positive d.c. voltage) disconnect the signal, or set the input coupling switch to ground, and use the vertical position control to align the trace with the bottom graticule baseline. If the trace went downwards, disconnect or ground the input and set trace to the top line.

Now reconnect the signal. Turn the attenuator clockwise as far as possible *before* the waveform goes off the screen. Now measure the offset from the reference line in divisions or centimetres (see Fig. 8.1a). Multiply this measurement offset value by the sensitivity of the attenuator and you have the d.c. component of the signal. In Fig. 8.1a the trace moved upward, so the reference level (with no input) was set to a low graticule line. The offset was 4.6 cm. The attenuator was set to (say) 10 V/cm, so the d.c. component was:

$$4.6 \times 10 = 46 \text{ volts d.c.}$$

In Fig. 8.1b, the downward movement of the trace indicates a negative d.c. voltage, so the reference line was set to the top. When the signal was reconnected, the trace moved down 6.7 cm, so the d.c. content was:

$$6.7 \times 10 = -67 \text{ volts d.c.}$$

Now let us go back to Fig. 8.1a. Set the input coupling switch to a.c. The d.c. component of the signal is now blocked by the input coupling capacitor. Now turn the attenuator further clockwise to give the maximum displayed signal within the screen area (see Fig. 8.2).

Let us assume that the signal at the +46 V d.c. potential is a sinewave.

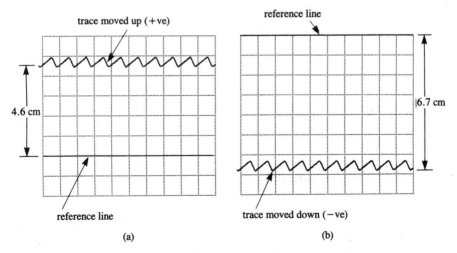

Figure 8.1. Screen displays when small a.c. signals with large d.c. offsets are measured.

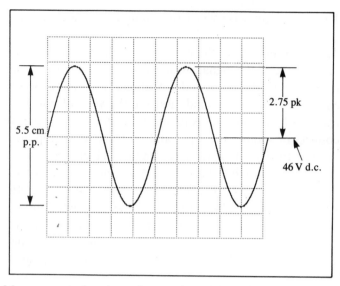

Figure 8.2. Measurement of peak, peak to peak, and d.c. offset values of a signal.

Carefully align the bottom tips of the waveform on a main graticule line and measure the overall height of the signal to the top tips.

8.1.2   Peak to peak

Now this overall height measurement is known as the peak-to-peak measurement. In this case (Fig. 8.2) it is 5.5 divisions. With the attenuator now set at 1 V/cm (say) the signal displayed is thus:

$$5.5 \times 1 = 5.5\,\text{V peak to peak (p.p.)}$$

So to obtain the peak-to-peak value of a waveform, measure from the most positive point to the most negative point within one complete cycle of the waveform.

8.1.3   Peak value

Where a waveform is symmetrical, i.e. the positive half-cycle is identical but of opposite polarity to the negative half-cycle (mirror image), the peak value is simply half the peak-to-peak value (see Fig. 8.2).

8.1.4   Root mean square (r.m.s.)

The r.m.s. value of a voltage is the amount of a.c. voltage in the waveform which would produce the same power dissipation in a resistor as an equivalent d.c. voltage. For example, if a 100 V d.c. voltage was applied across a 10 kohm resistor, the power dissipated in that resistor would be:

$$\text{Power} = \frac{\text{Voltage} \times \text{Voltage}}{\text{Resistance}} = \frac{100 \times 100}{10\,000} = 1 \text{ watt}$$

An a.c. voltage of 100 V r.m.s. will also dissipate 1 watt in the 10 kohm resistor as above. Now the r.m.s. value of a sinewave is 0.707 of the peak value. So 100 V r.m.s. divided by 0.707 will give the peak value. And the peak value multiplied by 2 will give the peak-to-peak value.

$$\frac{100}{0.707} = 141 \text{ volts}$$

$$141 \times 2 = 282 \text{ volts}$$

So a sinewave of 282 V p.p. would dissipate the same power in a resistor as 100 V d.c.

To repeat the equation from the other direction:

$$282 \text{ V p.p. divided by } 2 = 141 \text{ V peak}$$

$$141 \times 0.707 = 100 \text{ V r.m.s.}$$

It is interesting to note that the UK domestic mains voltage of 240 V is the r.m.s. voltage. So beware, the voltage supplied on domestic mains sockets is

$$\frac{240}{0.707} \times 2 = 679 \text{ volts peak to peak!}$$

The relationship between the r.m.s. value and the peak value, is 0.707 only in the case of a sinewave. For all other waveform shapes, the relationship is more complex, and other types of instrument are more suitable for determining the r.m.s. content of a signal.

There is another way to measure the peak-to-peak amplitude of a signal which is often quicker and easier and may be more accurate. The problem is the alignment of the maximum positive and negative peaks of the waveform with the graticule lines. Only the centre graticule line normally has minor graduations marked on it at 1 or 2 mm intervals, against which you can measure the signal. So this means using the X position control to alternately align the top and bottom peaks of the waveform against this centre line.

The easier method is simply to turn the timebase switch counterclockwise to a slower speed, perhaps by four or five switch positions, so that a raster display is produced (see Fig. 8.3). Now with a band of signal passing across the centre line the top and bottom limits of the display can be measured against the minor graduations on the centre scale. The peak-to-peak measurement is 4.6 divisions as before, so the waveform is verified as 46 volts.

This method will only work well for signals above about 500 Hz, since lower frequency measurements will result in a flickering display, rather than a solid band of signal. So for lower frequency measurements, it is easier to use the first method outlined above.

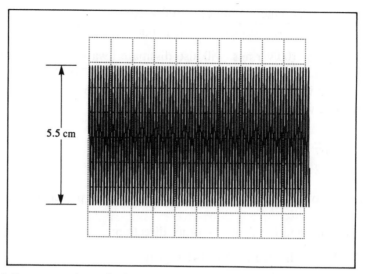

Figure 8.3. Raster waveform display for ease of peak-to-peak measurement.

8.2 Horizontal measurements

Before making measurements, the first general rule applies as before: make sure that variable (time) controls are set to the calibrated (CAL) position.

There are two main sections into which we can divide time measurements: time interval and period.

- A time interval is the distance between two points within one cycle or several cycles of a waveform.
- Period is the specific time between identical points on successive cycles of a repetitive waveform. Since frequency is the reciprocal (1 divided by) of time, the frequency of a waveform can thus be determined as follows:

$$\text{frequency } (f) = \frac{1}{\text{time (period)}}$$

Look at Fig. 8.4.

The interval between two repeating points (the period) of the waveform in Fig. 8.4 is 6 divisions (T_p). Therefore with the timebase set to $10\,\mu\text{s/div}$ (say), the period is

$$6 \text{ divisions} \times 10 \text{ (microseconds)}$$
$$= 6 \times 10 \times 10^{-6}$$
$$= 60 \times 10^{-6}$$
$$= 60\,\mu\text{s}$$

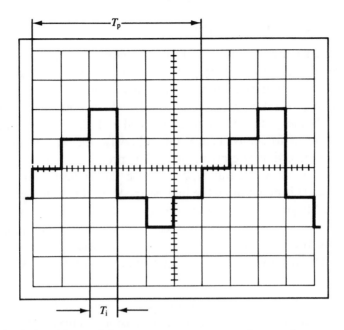

Figure 8.4. Period and interval time measurements on a pulse waveform.

Therefore the frequency of this signal is

$$\frac{1}{T_p} = \frac{1}{60 \times 10^{-6}}$$
$$= 1.66 \times 10^4 = 16.6 \text{ kHz}$$

To make a time measurement, it is easiest to align the first reference point on a graticule line using the horizontal position control, then count along the graticule to the right. The distance in divisions or centimetres between these two points is then multiplied by the timebase setting (CAL position).

So in Fig. 8.4, suppose we wish to measure the duration of the biggest pulse section of the waveform. The left-hand side of this pulse is aligned with a graticule line and the distance read off to the right-hand edge of the pulse (T_i).

In this case T_i measures 1.0 division. Therefore

$$T_i = 1.0 \times 10 \times 10^{-6}$$
$$= 10 \times 10^{-6}$$
$$= 10 \ \mu s$$

So the top pulse is $10 \ \mu s$ wide.

Now we come to a particular case involving careful use of the X and Y controls.

8.2.1   Risetime measurements

Usually with a square or pulse waveform, the risetime is the time taken to go from one steady state to another – say, from its negative level to its positive level or vice versa. Now risetime measurements are made in a particular way. The measurement is taken between the 10 and 90 per cent steady-state values (see Fig. 8.5). Some graticules are ready marked with the 0, 10, 90 and 100 per cent vertical levels for convenience, but otherwise the waveform has to be positioned on the ordinary graticule lines. The procedure is as follows.

In order to measure accurately, try to use the highest calibrated sweep speed possible. This may not be the highest speed on the oscilloscope. As well as seeing the 'diagonal' rise of the waveform, it is necessary to 'fit' on the screen the lower and upper steady levels, and too fast a sweep speed may result in only the diagonal rise being visible. So the highest sweep speed possible is where the lower steady level, upper steady level and risetime are all just visible on the screen.

1 Adjust the vertical display size using the attenuator and (Y) variable gain controls given an exact height of 5.0 divisions between the lower and upper steady levels.

2 Now use the vertical position control to set the bottom steady level 5 mm (0.5 div) *below* a horizontal line (H1). Now the 10 per cent level of the

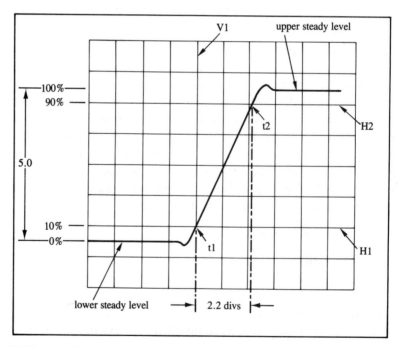

Figure 8.5. Points of measurement for determination of pulse risetime.

risetime is obviously 5 mm *above* the bottom steady level, so this 10 per cent point is now on a main horizontal graticule line.

3 Next use the HORIZONTAL position control to align the 10 per cent point with a VERTICAL graticule line, (V1) and note which one it is. The 10 per cent level will now be at the intersection of main horizontal (H1) and vertical (V1) graticule lines. This intersection of H1 and V1 will then give the first time reference point t1.

4 Now the 90 per cent level (H2) will be exactly 4.0 divisions above the 10 per cent level H1. Since the total waveform display height is 5.0 divisions, the 10–90 per cent points will thus be 80 per cent of 5.0 divisions, which equals 4.0 divisions. So measure up the graticule 4.0 divisions from H1 and read along to the right to the point where the waveform crosses the 90 per cent line H2. This point will not necessarily be on an intersecting vertical line.

5 Now measure the horizontal distance from the VERTICAL line t1 noted above, across to the point where the waveform crosses the 90 per cent HORIZONTAL line t2. This distance, or time interval, between t1 and t2, is thus the 10–90 per cent time, or risetime of the pulse.

Now multiply this measurement by the timebase setting to obtain the risetime of the pulse. Make sure the timebase VARiable is still set to the CALibrated position.

In Fig. 8.5, the distance $t2 - t1 = 2.2$ divisions.

The timebase speed is (say) 50 ns/div.

So the risetime $T_r = 2.2 \times 50 \times 10^{-9}$

$$= 100 \text{ ns (nanoseconds)}$$

Having established the technique for risetime measurement, there are two very important factors to consider before interpreting the measured risetime value. These are the risetime of the oscilloscope being used to make the measurement, and the risetime of the probe you are using. The measured risetime will be a combination of the risetimes of (i) the original signal, (ii) the probe in use and (iii) the measuring oscilloscope. If the scope and probe risetimes are small compared to the measured value (say one-tenth or less), they can be ignored for practical purposes. But if the scope and/or probe risetimes are greater than one-tenth of the measured value, then the actual signal risetime must be calculated. If the scope and/or probe risetimes are not known, they can be found from the following relationship:

$$\text{Risetime (ns) } T_r = \frac{350}{\text{Bandwidth } (f) \text{ in MHz}}$$

So, for example, if the scope bandwidth is 50 MHz, its risetime is

$$T_r = \frac{350}{50} = 7 \text{ ns (7 nanoseconds)} = 7 \times 10^{-9} \text{ seconds}$$

Suppose the probe has a 100 MHz bandwidth, then the probe risetime is

$$T_r = \frac{350}{100} = 3.5 \text{ ns}$$

Let us call the signal risetime that we are going to measure T_r, the oscilloscope risetime T_o, the probe risetime T_p and the combined risetime (measured) T_c, then

$$T_r^2 = T_c^2 - T_o^2 - T_p^2$$

Therefore

$$T_r = \sqrt{T_c^2 - T_o^2 - T_p^2}$$

Thus in this example, the signal risetime

$$T_r = \sqrt{110^2 - 7^2 - 3.5^2}$$

Thus

$$T_r = \sqrt{12\,100 - 49 - 12.25}$$

Thus

$$T_r = \sqrt{12\,038.75}$$

Hence

$$T_r = 109.72 \text{ ns}$$

So here where the probe and scope risetimes are small compared to the measured total, the exact value 109.72 only varies from the (approximate) measured value by about 0.3 ns in 110 ns – much less than 1 per cent. So here the measured value is good enough.

If the scope and probe were about one-tenth of the measured total, i.e. 11 ns (scope), 11 ns (probe) 110 ns (total) then

$$T_r = \sqrt{110^2 - 11^2 - 11^2}$$

So $T_r = 108.89$ and the error is just about 1 per cent, so this is about the point where the errors must be considered.

If the risetime being measured was fast, and the observed value was 10 ns, then using the same scope and probe as before ($T_r = 7$ ns, $T_r = 3.5$ ns) then

$$T_r = \sqrt{10^2 - 7^2 - 3.5^2}$$
$$T_r = 6.2 \text{ ns}$$

So here there is a very significant difference between the measured value (10 ns) and the actual signal risetime of 6.2 ns. You can see that in this case with fast risetimes, the measured value of 10 ns cannot be taken as the result. The calculated value of 6.2 ns is the correct result. The error is so large (38 per cent) that the calculation must be done for accurate results.

8.2.2   Delay timebase

Some oscilloscopes feature a dual timebase or delay timebase system. This is particularly useful when examining a complex waveform which can only be triggered at a particular point. On a simple repetitive waveform such as a sinewave or square wave, the whole of one cycle can usually be expanded, and by the use of the trigger level and polarity controls, the trigger point can be varied so as to start the timebase just at the point of interest (or just before) on the waveform. Once the vital part is displayed on the left of the screen, the timebase speed and the horizontal magnifier can be used to carefully examine the detail.

However, with a complex waveform it is often impossible to trigger on the point of interest. So if the vital part is not on the left of the display, increasing the timebase speed will only 'push' the part you want to see off the right of the screen. To overcome this problem, the scope is triggered where possible on the waveform, as near as possible before the point of interest. Then a (variable) time delay is introduced (see Fig. 8.6). This delay time is equal to the distance (in time) between the timebase trigger point and the point of interest. At the end of this delay period, the delay 'timer' provides a pulse which then starts a second timebase (usually B) or restarts the main timebase. Now with the sweep starting just at the vital point of interest, the sweep rate can be increased to magnify the waveform and carefully examine the content.

Although for clarity Fig. 8.6 shows a magnification of about 10:1 of the detail of the waveform, in practice, magnifications of at least 1000:1 are possible. At these high delay ratios, it is essential to have a very bright CRT and invariably a PDA type CRT is used with an overall accelerating voltage of at least 10 kV.

8.2.3   After delay trigger

A further benefit on many oscilloscopes with a delay timebase, is the after delay trigger facility. Looking carefully at the signal waveform illustrated in Fig. 8.6, you can see that after the delay time, the B sweep immediately starts, and the delayed portion is displayed (expanded) on the screen. Now there can be a few problems with this facility mainly due to electrical 'noise'. At very high sweep delay ratios, that is where the delay time period is large compared with the delayed sweep time, any amount of electrical disturbance in the delay

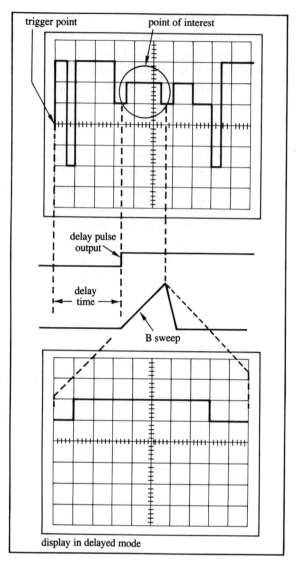

Figure 8.6. Use of a delay timebase to expand a horizontal section of a waveform for close examination.

timebase system will cause the displayed waveform to 'jitter'. This is observed as intermittent left and right movement of the waveform of interest on the screen in the delayed mode.

Now provided that the delay time was adjusted so as to start the 'B' sweep just before the point of interest, we can now utilize a second trigger system: the 'after delay' or 'B' trigger system. The vertical amplifier system is connected to the B trigger circuit just the same as to the main A trigger system. So each time the input signal passes through the trigger threshold, a sharp output pulse is produced. Referring to the same waveform in Fig. 8.6, look at the output from the B trigger circuit, with the trigger level control B adjusted to a threshold level as shown (see Fig. 8.7). It can be seen that there is a trigger edge at the point of interest on the waveform.

So now, instead of starting the B sweep after the delay time, the delay circuit pulse 'enables' the B sweep, or allows the B sweep to start when triggered. However, the B sweep does not start until a short time later when it is 'fired' by the sharp pulse from the B trigger circuit.

Since the B trigger point is precisely time locked to a point on the input waveform (just as in the normal 'A' or single timebase triggering system) there is no jitter apparent on the screen. So here you can see the great benefit of observing a small portion of a signal magnified from a much bigger signal, solidly displayed on the screen.

8.2.4   *Phase measurements*

There are two ways to measure the phase difference between two signals on an oscilloscope. One way is to use the dual trace mode, and the other is to use the X–Y mode. For the X–Y method to work, the X and Y inputs, usually utilizing the CH1 and CH2 vertical inputs, must have equal frequency and phase response at the frequency of the signals you want to measure.

Firstly, couple the two signals to the CH1 and CH2 inputs and adjust the timebase switch to display both signals clearly.

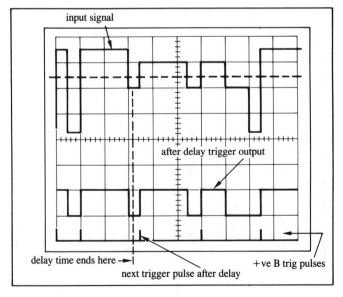

Figure 8.7. The time relationship between the input signal of Fig. 8.6; and after delay trigger pulses.

X–Y phase method

Now adjust CH1 and CH2 attenuator and variable controls so that each channel displays *exactly* equal amplitudes of a convenient height, say 6 divisions (see Fig. 8.8). Now select X–Y mode. The shape of the display will vary from a straight line (0^0 or zero phase shift) through an elliptical shape, to a circle (90^0 phase shift). We will assume here that the phase shift is somewhere in the middle.

Note that Figs. 8.8 and 8.9 are both displays of the same two signals, first in the dual trace mode, and then in the X–Y mode.

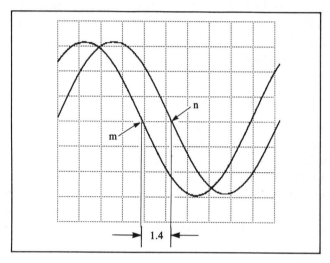

Figure 8.8. Diagram showing two sinewaves, and measurement points for phase differences using the dual trace method.

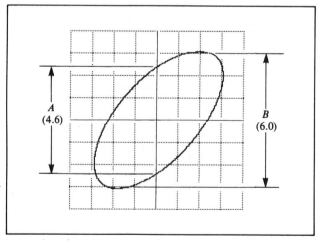

Figure 8.9. X–Y display of two sinewaves with measurement points of major and minor axes.

Now the ellipse (Fig. 8.10) must be positioned centrally on the screen. Use the position controls to set the display so that the distance (a to o) equals (o to b); and (c to o) equals (o to d). Now the phase angle can be calculated from

$$\sin \phi = \frac{A}{B}$$

where A and B are the diameters of the minor and major axes of the ellipse (Fig. 8.9).

Measure A as the distance between the two points where the ellipse crosses the horizontal centre line: here 4.6 divisions.

Measure B as the total vertical height of the display: here 6 divisions.

$$\text{so } \sin \phi = \frac{4.6}{6.0}$$
$$= 0.76$$

Thus ϕ = the angle whose sin = 0.76 or ϕ = arcsin 0.76
From sin tables or a calculator, arcsin 0.76 = 50°.
So the phase angle between the two signals is 50°.

Dual trace method

As before the two signals are connected to the CH1 and CH2 inputs, both adjusted to give exactly equal vertical display heights (here 6.0 divisions). Now adjust the trigger level control (on manual trigger) to set the trigger point half-way up (here on the vertical centre), and adjust the timebase switch and variable so that one complete cycle takes exactly 10 divisions.

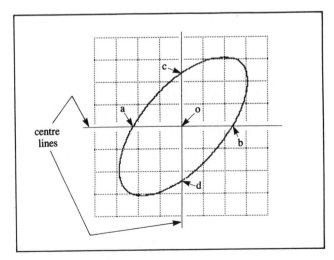

Figure 8.10. Diagram showing X–Y display of two sinewaves indicating reference points of the ellipse on the graticule.

Now one cycle is 360°, so 360/10 = 36° per division.

Measure the distance m–n (Fig. 8.8) between two crossing points on the vertical centre line (here 1.4 divisions).

So 1.4 × 36 = 50° phase difference (as before).

8.2.5 *Amplitude modulation measurements*

It is quite simple to use the oscilloscope to display and measure amplitude modulation. Figure 8.11 shows a display of a typical amplitude modulated carrier wave.

The (higher frequency) carrier wave has its peak-to-peak amplitude varied (modulated) by the lower frequency, modulating sinewave. In practice, the carrier frequency would be much higher in relation to the modulation frequency, and the modulating signal would probably not be a simple sine-wave. When looking at a few cycles of the (lower frequency) modulating signal, it would not normally be possible to discern the individual cycles of the carrier signal. The modulating waveform may be varying harmonics such as in speech, digital pulses such as in radio control systems, or almost any type of signal. To measure the modulation depth it is necessary to measure the maximum and minimum peak-to-peak levels of the carrier signal, caused by the modulation signal. Because of the difficulty of triggering a mixed frequency waveform such as this, and then determining the maximum and

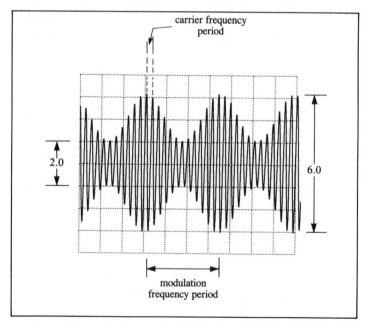

Figure 8.11. Normal timebase display of an amplitude modulated sinewave.

minimum points, it is convenient to use the oscilloscope in the X–Y mode, so either an external X input socket is required, or preferably, the ability to use two vertical amplifiers on a dual trace instrument as matched X–Y inputs.

The a.m. signal consists of a high frequency carrier signal, and a lower frequency modulating signal, such as speech.

The modulated carrier is fed into the vertical input with the oscilloscope in the X–Y mode. The horizontal (X) scanning signal is provided by the modulating signal which is thus connected to the X input. The input attenuators and variables are then adjusted to give a convenient size display on the screen, as large as possible within the screen area. This is one occasion when the variables do not affect the measurement, since it is the relative values that are measured. The resultant waveform is a display of the modulation depth (see Fig. 8.12).

The percentage depth of modulation is given by

$$\frac{A - B}{A + B} \times 100 \text{ per cent}$$

So if $A = 6.0$ divisions and $B = 2.0$ divisions, the modulation depth is

$$\frac{6 - 2}{6 + 2} \times 100 \text{ per cent} = \frac{4}{8} \times 100 \text{ per cent} = 50 \text{ per cent}$$

Note that Figs. 8.11 and 8.12 both show the same signal, in the YT and XY modes respectively.

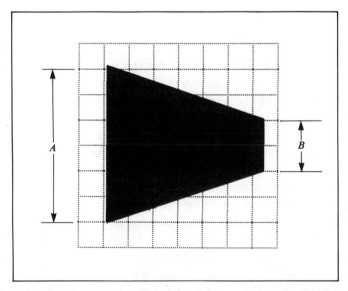

Figure 8.12. X–Y display of amplitude modulated sinewave, showing points of measurement for determination of modulation depth.

When the right-hand side (B) is reduced to zero, the modulation is 100 per cent, i.e.

$$\frac{6-0}{6+0} = 1 \times 100 \text{ per cent} = 100 \text{ per cent}$$

Excessive levels of modulation input cause distortion and this can be seen as non-linearity, distortion and zero amplitude (straight line) on the display.

8.2.6 Cursor measurements

An increasingly popular feature on modern oscilloscopes is the cursor measurement facility. Circuitry within the oscilloscope is used to generate a pair of dotted or dashed lines on the CRT. This pair of lines can then be selected to appear in the vertical or horizontal axes for horizontal or vertical measurements respectively. The position of each of the lines is independently adjustable, and then the distance between them is used to calculate the time or voltage difference between them. These calculations are made by the oscilloscope and the results are displayed in numeric format for direct reading. The Greek symbol delta (Δ) is used to denote the functions for voltage or time measurements. This is derived from the mathematical expression Δx, meaning 'change in x'. So Δt selects the horizontal mode for measuring the *change in time* between the two cursors, and Δv selects the vertical mode for measuring the *change in voltage* between the two cursors. So the symbols Δt and Δv are used for the front panel coding of the cursor mode selection switch.

Vertical cursor measurements

First the cursor lines must be selected for the vertical measurement mode. This means selecting the Δv mode on the selector switch. The two horizontal cursor lines displayed on the screen may be identified as lower and upper or 1 and 2. This might be by a difference in their appearance on the screen, for instance a dotted line for the lower (1), and a dashed line for the upper (2). Alternatively (or perhaps additionally), they might be identified by the labelled position controls that move them. The voltage difference between the two lines is then displayed either on the CRT as a screen readout display, or on the front panel. In order to determine the voltage magnitude between the two cursor lines, it does not matter which cursor line is set as the top limit and which is the bottom. However, in order to determine the correct polarity of the voltage measured, the identified upper cursor line should be used as such and placed at the top of the waveform, and the lower line at the bottom. Then the difference between the lower cursor line and the more positive, upper cursor line will be displayed as a positive measurement, e.g. +2.4 V. Similarly, measuring the voltage difference between the lower cursor line and the (upper) cursor line placed below it will result in a negative voltage readout, e.g. −1.7 V. This facility is useful is useful for measuring non-repetitive or d.c. voltages.

The method of vertical (voltage) measurement is the opposite of normal graticule measurements. Instead of the waveform being placed on a horizontal line, and measured off against another horizontal graticule line, the waveform is kept still and the cursor lines moved to the top and bottom of it.

As with all measurements, first ensure that the VARiable gain control is set to the CALibrated position. This is often indicated on the screen readout display, e.g. 'UNCAL', when in the uncalibrated condition.

Peak-to-peak measurement

Now using the cursor position controls, set the lower and upper to the bottom and top limits of the part of the waveform you want to measure. In the case of peak-to-peak measurement, this will be the maximum bottom limit of the waveform, to the maximum top limit (see Fig. 8.13).

Very accurate measurements can be made using the cursor system since the lines can be precisely adjusted to the waveform peaks and the voltage value directly read. With the normal graticule method, a degree of estimation is required to measure a waveform peak against the limited number of minor graduations on the graticule lines.

It is important to note that the displayed voltage magnitude does *not* make allowance for any probe attenuation factor. So although the screen readout may indicate a voltage of, say, 670 mV, if you are using a ×10 probe, this

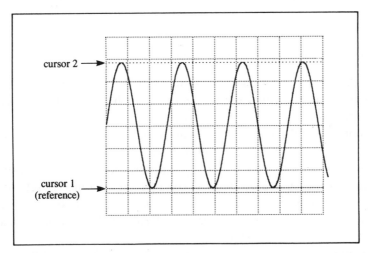

Figure 8.13. Diagram showing cursor alignment for measurement of the peak-to-peak amplitude of a signal.

value must be multiplied by 10 to obtain the true voltage at the probe tip. So in this case the signal voltage would be $0.670 \times 10 = 6.7\,\text{V}$. Remember that with a $\times 10$ attenuating probe, the voltage reaching the oscilloscope input will be 10 times smaller than at the probe tip; or the probe tip voltage is 10 times greater than the voltage at the oscilloscope input.

If you are using a more sophisticated oscilloscope, the probe factor may be dealt with automatically by the oscilloscope and manual calculations may not be necessary (provided the manufacturer's probes are used). However, in most cases it will be necessary to make the calculations yourself. Thus, since the probe tip voltage is 10 times larger than the displayed voltage measurement, *multiply* this value by 10 to obtain the correct signal voltage.

Obviously if other attenuation probes are used, such as $\times 100$, then this factor must be used to multiply the display voltage magnitude, i.e. probe tip voltage = displayed voltage $\times$ 100.

Partial waveform measurement

It may be necessary to measure between two vertical points on a waveform which are not the top and bottom peaks. Again use the *lower* (1) cursor line as the bottom or reference line, and position the *upper* (2) above it. Usually the voltage difference can be measured between any two points within the graticule area, so it is simply a matter of positioning the lower cursor line on the lowest point you want to measure from, and the upper line to the highest point. Then simply read off the voltage difference from the screen readout. The scaling factor is computed within the oscilloscope, so there is no need to worry which attenuator range you are using. Once again, the only potential problem is the probe dividing factor, as described in the peak-to-peak section above.

Direct current (d.c.) voltage measurements

If you wish to use your oscilloscope to measure d.c. voltages, the cursors can be particularly helpful. First use the *lower* cursor line (1) as the reference for the measurement.

If the polarity of the voltages you want to measure is always the same, you can set the reference level to the top or bottom of the screen accordingly to maximize the visible deflection when the input is connected. For instance, if you intend to measure only *positive* voltages on Channel 1, set the input coupling switch to ground, then use the Channel 1 vertical position control to set the trace to the bottom graticule line. Then set the *lower* (1) cursor line also on the bottom graticule line.

If you intend to measure only *negative* voltages, first set the 'grounded' trace and *lower* (1) cursor line to the top graticule line.

Next set the Channel 1 input coupling switch to d.c., and connect your probe to the d.c. voltage. Adjust the Channel 1 attenuator for maximum deflection within the graticule area, then set the *upper* (2) cursor on the (deflected) Channel 1 trace line. The readout will now display the correct polarity and magnitude of the d.c. input voltage.

If the intended measurements might be positive and negative d.c. voltages, then before connecting your probe to the d.c. voltage, set the 'grounded' trace and the *lower* (1) cursor line to the vertical centre graticule line.

Now once again, any voltage connected to the oscilloscope will be indicated with both polarity and magnitude on the readout display once the cursor 2 line has been set on the deflected trace.

Remember to multiply the indicated voltage measurement by the probe attenuation factor to obtain the true value of the voltage at the probe tip.

Horizontal cursor measurements

Select the horizontal or time measurement mode Δt on the cursor selector, so that the two vertical cursor lines are displayed on the screen. The left-hand (1) cursor is then positioned at the first part (in time) of the waveform from which you wish to make your time measurement. The right-hand cursor (2) is then positioned further to the right to the end of your chosen time interval. If it is the period and/or the frequency of the waveform that you want to measure, then the two cursor lines must be positioned at the beginning and end of *exactly* one cycle of the waveform.

Period time measurement

In order to be able to measure the waveform period, at least one cycle of the signal must be displayed on the screen. For the most accurate measurement, the minimum number of cycles should be displayed, so set the timebase switch to the fastest speed possible, while still displaying at least one full signal cycle on the screen. Set the time VARiable control to the CALibrated position. Now using the cursor 1 position control, set the left cursor to an exact point at the beginning of the waveform cycle. In the case of waveforms with sharp fast rising edges, this could conveniently be on such a sharp rising edge. With slow rising waveforms, such as sinewaves, it is easier to first set the waveform equally above and below a horizontal graticule line, preferably the centre line. So first use the vertical position control to set the trace equally about the vertical centre line.

A convenient way to do this is to adjust the waveform amplitude to an exact whole number of divisions such as 4.0 or 6.0. You can do this by adjusting the attenuator and Y VARiable controls. You can then set the vertical position control to give 2.0 divisions above and 2.0 below the centre line (or 3.0 above and 3.0 below with a 6 division display). Now set the cursor 1 line to the point on the left where the waveform first crosses the centre line (see Fig. 8.14).

Set the cursor 2 line to the point on the waveform where, after one full cycle, the waveform repeats, and again crosses the centre line. Now read the Δt time difference on the screen readout. During this measurement process, ignore the vertical graticule lines normally used for time measurement. The only graticule line of interest here is that horizontal line passing through the vertical centre of the waveform.

Now that the period of the waveform has been measured, the oscilloscope will usually have the facility to compute the frequency. This is simply a matter of selecting the frequency or (f) function in the cursor area of the front panel, to see the waveform frequency directly displayed on the readout. If this facility is not available, then the frequency can be calculated from frequency (f) = $1/t$. So in that case, calculate the reciprocal of the displayed Δt time measurement, i.e. divide 1 by Δt.

Cursor time measurement

Whether or not the waveform displayed is a periodic repetitive signal, you can make time measurements over any part of the waveform. For instance if you have a pulse waveform and you wish to check the mark to space ratio, you can do this by measuring first one 'half' cycle and then the other. Strictly

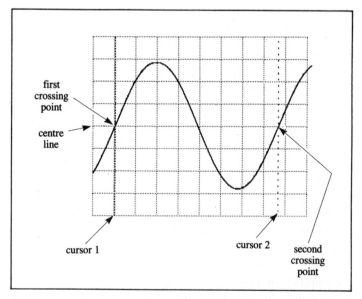

Figure 8.14. Diagram showing cursor alignment for measurement of the period time of a signal.

speaking, if the duty cycle of the waveform is not 1 : 1, then you cannot really call the first section a 'half' cycle. However, in practice on a simple repetitive signal, the meaning is clear.

Select cursor 1 as before and position the line at the beginning of the pulse. Select cursor 2 and position the line at the end of the pulse. Now read off the time from the display (see Fig. 8.15).

Now select cursor 1 again, and move the line to the right to the end of the second pulse, and read off this second displayed time value. Now these two measured times give the mark to space ratio of the signal. If the duty cycle is 1 : 1, such as with a square wave, then this ratio can be checked with another facility usually found in association with cursor measurement namely the cursor tracking system.

Cursor tracking

The tracking facility allows both cursors to be moved simultaneously by one control. So once a fixed time or voltage interval has been set between the cursor lines, the distance or interval is maintained while the two lines are moved across the screen.

In the case of time measurement, referring to the example above concerning duty cycle measurement, the cursor 1 line can be positioned at the beginning of the pulse and cursor 2 at the end. Then by introducing the tracking mode,

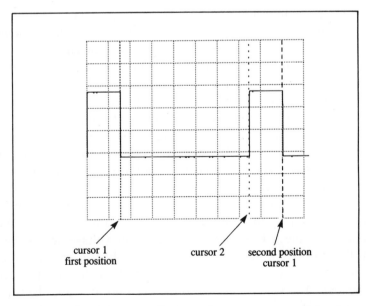

cursor 1
first position   cursor 2   second position
                                                 cursor 1

Figure 8.15. Diagram showing cursor alignment for measurement of time interval of a signal.

both cursors can be moved across to the next pulse to see that it has exactly the same duration as the first.

If the waveform is complex, then both cursors can be moved together along the signal to check that several sections have the same duration.

Also the interval between the cursors can be set to a precise time by watching the time readout as cursor 2 is moved relative to cursor 1. For instance, the time between the cursors could be set to exactly 5.00 ms, then the tracking mode engaged, and this 5.00 ms interval moved across the screen to check that various elements of a complex waveform are all 5.00 ms wide.

The tracking system can equally be used in the vertical cursor mode to check the amplitude of different voltage levels in a waveform. With a staircase waveform for instance, the pair of cursor lines can be moved up each step of the waveform to check for equal stair height. Again, this can be after first setting the cursor interval to one of the step amplitudes, or calibrating to a fixed exact voltage increment by setting against the readout display.

APPLICATION EXAMPLE

To examine the contact bounce of a switch

In this example we shall use the store, single sweep, save, pretrigger and cursor measurement functions to measure the contact bounce on a switch. When a standard, low cost switch is operated, the contacts do not make immediate, abrupt, continuous contact. That is, they do not immediately switch from open circuit to short circuit connections. For most applications this effect is not a problem, but in the case where there may be a problem, it can easily be eliminated by the fitting of a small capacitor. In order to determine the correct value of this capacitor, it is essential to know the nature and duration of the 'bounce' effect. In practice the contacts tend to 'bounce', causing intermittent contact and generating a small voltage transient before remaining fully closed. To study this effect, a digital storage oscilloscope is used to capture this contact bounce in memory, then measure its duration and amplitude using the cursor measurement system.

The circuit is connected as in Fig. 8.16. A standard 1.5 volt battery is used to supply the voltage to be connected by the switch. With the battery connected as shown, the resultant voltage across the probe will be positive with respect to ground, so the trigger polarity selector should be set to the positive slope mode. A 1 kohm resistor is used as the load. The switch S1 is set to the open position and the Channel 1 probe set to the ×10 position, then connected across the switch as shown in the diagram.

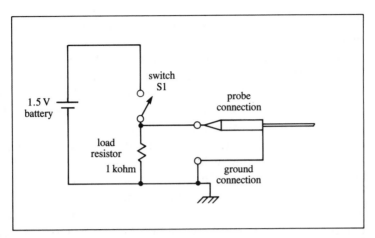

Figure 8.16. Diagram showing the probe connections to a switch for examination of contact bounce.

Since the amplitude and duration of the expected voltage transient is unknown, several attempts may be required to find the optimum timebase, trigger level and attenuator settings. If the oscilloscope has a pretrigger facility, this can be used so that when the signal is captured, the display starts at a point in time, say 25 per cent before the transient occurs, that is 25 per cent of one sweep duration time.

If there is no pretrigger, then the trigger level must be set carefully so that when the signal occurs, the oscilloscope timebase is immediately started, so as not to miss part of the waveform. This can be achieved by first carefully setting the trigger level control point with a continuous signal.

A 1 kHz square wave such as from the calibrator socket would be suitable, or any sinewave signal above 50 Hz. Adjust the Channel 1 attenuator to give the minimum triggerable display amplitude specified for your oscilloscope (usually about 5 mm). Taking this to be the case, with the 0.5 division display height, now select manual trigger. Then adjust the LEVEL control to give a stable triggered display of the signal. Do not then alter the setting of the LEVEL control as it is now set to the optimum position for the switch test. Disconnect the calibrator signal from the Channel 1 input socket and connect the probe as in the circuit diagram (Fig. 8.16).

First select Channel 1 only on the oscilloscope and adjust the Channel 1 attenuator to 50 mV/div and select manual trigger from Channel 1 source. (Do not alter the LEVEL control.) With the ×10 probe we will have a sensitivity at the probe tip of 0.5 V/div, so the 1.5 volt cell should give us a positive going step display of about three divisions. Try repeatedly operating the switch S1 and check that the timebase is triggered every time.

Now select STORE mode, and also select SINGLE sweep, then press the RESET button. Once 'armed' the single sweep condition is normally indicated by an illuminated lamp on the front panel. The probe should now be connected between the Channel 1 input socket and the switch S1, as shown in the diagram.

It is now necessary to experiment with the oscilloscope timebase setting to capture the switch waveform for the best display of its characteristics. Close switch S1, and observe the effects on the oscilloscope; the trigger lamp should flash, the single shot lamp should go out, and the screen should now display a stored waveform. Now the objective is to try to get the waveform as wide as possible across the screen while still showing the complete signal event. So after the first event capture, check the display and see how to improve it. The timebase may need to go faster for better waveform interpretation, or slower to contract the signal width to fit the whole event within one sweep time. If necessary adjust the timebase switch one step in the required direction, open the switch S1, and press the SINGLE sweep RESET button.

Now once again close the switch S1 and check the new display. If it is not optimized, repeat the above procedure as many times as necessary until you obtain the best results possible.

If the oscilloscope has a pretrigger facility, this should be used and set to lowest position (above zero) such as 10 or 25 per cent before starting the signal capture sequence above. With pretrigger in use, the signal will be displayed about one division (10 per cent) or two-and-a-half divisions (25 per cent) from the left-hand edge of the graticule.

Once a signal has been captured with suitable proportions of height and width, secure this waveform in the memory by pressing the SAVE Channel 1 or HOLD Channel 1 button.

Note that in this example 25 per cent pretrigger was used and that the signal shown in this example is peculiar to the actual switch used, and that any other switch may give a completely different characteristic waveform. It can be seen that in this example there is a main group of 'bounces' before the voltage transfers from zero to the 1.5 volt level, then after the voltage has remained settled at the high level, there is a further small burst of 'bounces'. Firstly, we shall measure the time to the end of the first group of bounces, then later the total time from the operation of the switch until all the aberrations have finished.

For the time measurement select the Δt horizontal cursor mode. Adjust cursor 1 to the first point in time where the waveform starts. If

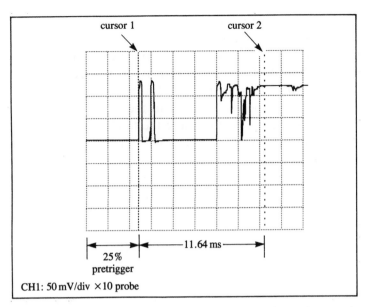

Figure 8.17. Contact bounce waveform with cursors aligned for peak-to-peak measurement.

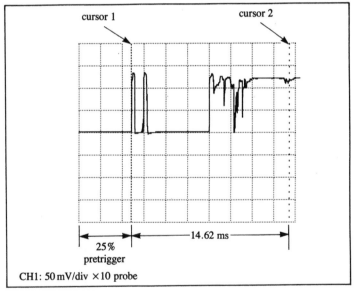

Figure 8.18. Contact bounce waveform with cursors aligned for duration time measurement.

you used pretrigger, then this start point will be the first deviation on the left from the straight line (baseline) level (in this case 25 per cent of 2.5 divisions from the left). Without pretrigger, the signal will start at the extreme left edge since the start of the signal will have caused the start of the sweep. Now adjust cursor 2 to the end of the burst of signals on the right. In this case it is set to the end of the first 'main' group of bounces. The duration of the switching transient can then be read out from the Δt readout display. In this example the duration was found to be 11.64 ms (see Fig. 8.17).

The cursor 2 line was then set to the right side of the last of the aberrations (Fig. 8.18) to measure the total bounce time from beginning to end. In this example the total time for the switch contacts to settle was measured as 14.62 ms.

9
Earth or ground connections

If you refer to the beginning of this book, you will see that oscilloscopes make measurements using a two-terminal system: one 'active' input to the Y amplifier, and one reference connection usually connected to the earth terminal or 'ground' system. In the simple case of the measurement of a battery voltage, since the battery is not connected to anything else, it does not matter which way round the connections go to the battery. The only difference is that if the positive terminal of the battery is connected to the input terminal of the oscilloscope, and the negative terminal connected to the ground lead of the instrument, the trace will be deflected upwards, and if the battery connections are reversed, the trace moves downwards. However, in practice, oscilloscopes are connected to many forms of electrical apparatus, not just a battery, so there are some very important things to consider.

9.1   Mains safety earth

Most oscilloscopes are supplied with their chassis, cabinet, input (ground) references and any exposed metal parts *connected to mains safety earth*. This is to ensure that the oscilloscope metalwork, which can be touched by the operator, remains at ground potential. If this connection were removed, it would be possible under certain conditions for the oscilloscope metalwork to be raised to a high potential relative to earth. Then an operator touching the instrument would be in extreme danger if he or she also touched some earthed object such as a water pipe, or other earthed apparatus. By keeping the oscilloscope chassis connected to earth, this danger is eliminated, so the *protective safety earth should never be removed*. However, not all oscilloscopes are supplied with earthed chassis systems. There are two alternatives available, although as stated above the grounded chassis system is the most common.

To meet safety standards in the United States oscilloscopes are manufactured to American safety standard UL1244. The European safety standard which includes oscilloscopes is IEC348. These standards stipulate many requirements to minimize the hazard risks. As well as various regulations regarding insulation, safety testing, minimum voltage breakdown and so on, there is a requirement for safety earth or ground protection. The way that the mains earth is connected (or not) and the subsequent insulation and isolation of the live and neutral connections of the mains fall into two categories. These are Safety Class I and Safety Class II. There are in fact four categories, 0, I, II and III, but only Classes I and II are applicable to standard oscilloscopes. Class 0 denotes no protection and Class III denotes special protection with special safety requirements and low voltages, so only Classes I and II apply to standard oscilloscopes. Safety Class II denotes inherently safe equipment without provision for protective earthing.

With Safety Class II equipment, the mains input connections are insulated from the instrument with a special transformer incorporating a flameproof insulation barrier. This is to ensure that under no circumstances, including failure of the transformer, can the mains connections become directly connected to any part of the instrument. So the primary, mains input side of the transformer is completely isolated from the secondary side supplying all the oscilloscope operating voltages (see Fig. 9.1). With this design, the 0 V reference (ground) points are *not* connected to earth. The chassis, 0 V reference and all accessible conductive parts are 'floating' with respect to earth. So within certain limitations, the 0 V reference of the oscilloscope can be connected to a potential which is not at ground level. Instruments manufactured to Safety Class II are tested with more than 1000 V between the instrument chassis, and the mains input connections.

Safety Class I is more commonly adopted as the safety standard for oscilloscopes, and hence they are supplied with safety protection ensured by

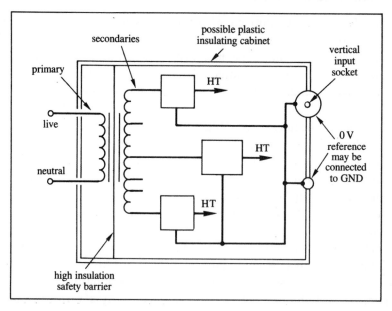

Figure 9.1. Safety Class II type equipment with isolated measurement reference.

special connection to the protective earth conductor. This must be achieved by *separate* connection, not forming part of any other circuit, with minimum specified sizes of connections and conductors. This connection must *not* be removed. Figure 9.2 shows the connection diagram for the Class I, earthed chassis system, and this should be compared with the Safety Class II system shown in Fig. 9.1.

9.2 'Signal earth connections'

When the instrument is used to observe low frequency, high amplitude signals, the type of earth used to connect the oscilloscope to the test apparatus is not too important. Most modern probes are supplied with an integral earth lead and clip, but at low frequencies it may be more convenient to use a separate long wire. This would be connected between the earth reference point (chassis?) of the equipment being checked and the oscilloscope ground. This saves constantly moving the earth clip every time the probe is moved to a different point, and works well for signals slower than a few kilohertz. However, above this frequency, only the short integral earth lead should be used. At high frequency the resistance and inductance of the earth lead itself start to become significant and cause apparent distortion to a viewed wave-form (Fig. 9.3). So for observation of high frequency signals, and as a general rule if possible, use the shortest earth lead you can. This will ensure a clean and accurate reproduction of the signal (Fig. 9.4). Note that Figs. 9.3 and 9.4

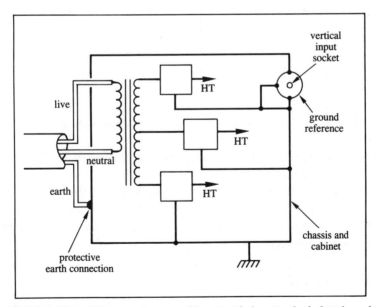

Figure 9.2. Safety Class I type equipment with grounded or earthed chassis and measurement reference.

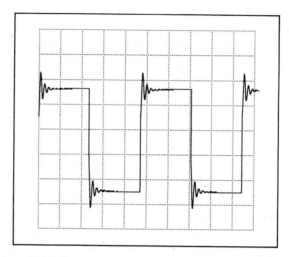

Figure 9.3. Display of high frequency square wave with poor probe earthing.

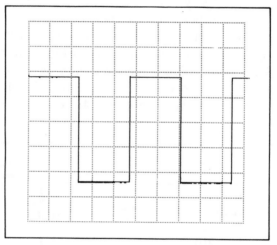

Figure 9.4. Display of high frequency square wave with good probe earthing.

show the same signal, with long and short probe earth leads respectively (1 MHz square wave).

The earth lead should be placed as close as possible to the part of the circuit you are examining. If two probes are used, with a dual trace instrument, each probe earth lead should be connected as close as possible to the nearest earth point for each probe respectively.

9.3   'Live' chassis

There are often circumstances where there is a problem in making the probe earth connection to the apparatus to be checked. This is when the chassis or reference level of the apparatus is not at ground potential. Probably most common of these is the television receiver, where the metal chassis is connected to one of the mains power connections, and may be live or neutral.

Under these circumstances the objective should not be to remove the protective safety earth from the oscilloscope, which would make it hazardous, but always to keep the oscilloscope earthed. The equipment should then be connected so that the chassis reference level can be connected to earth (via the oscilloscope ground lead). In the case of the television receiver, and in most circumstances, this can be achieved by the use of an isolating transformer. This is a mains transformer with an input to output voltage ratio of 1 : 1. So if the mains supply is connected to the input of this transformer, the same voltage appears at the output terminals. The power or VA rating of the transformer must be sufficient to supply the apparatus connected to its output, for example a television receiver.

Since there is no direct connection between the input and output terminals, either *output* terminal may be connected to earth (see Fig. 9.5).

So when the apparatus is connected to the output of the isolating transformer, the chassis or reference of that apparatus may now be connected to ground. This may be done permanently by a wire connection, directly between the chassis and earth (optional), or via the oscilloscope ground lead on the probe, which is already connected back to earth via the oscilloscope's own earthing system mentioned above.

If these steps cannot be taken due to the nature of the equipment being tested, then there are other methods available such as opto isolators for the oscilloscope signal input, and even special oscilloscopes with isolated inputs (active and reference connections). Professional TV service areas usually adopt a special isolated 'earth free' system, but this is not applicable to domestic use.

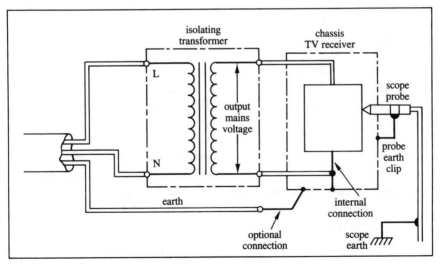

Figure 9.5. The use of a mains isolation transformer for safe operation of a TV receiver.

9.4   Earth (ground) connections

When there is more than one earth path in use while examining a signal on an oscilloscope, an interference signal may develop which then appears super-imposed on the signal. For instance on a dual trace instrument, when two probes are used and each of their ground leads is connected, such a signal may be seen on the waveform. It may be possible to remove the interference by connecting both the ground clips to exactly the same reference point, or sometimes better still to use only one of the ground leads. There is no fixed rule, but in general, for high frequency work use two very short ground references to a common point; and at low frequencies use just one ground reference connection.

Another situation causing earth loops is when the equipment you are testing has a (mains safety) earth connection (Safety Class I), and the oscillo-scope also has its own safety earth as discussed above. In this situation there are two alternatives. You should either use another different oscilloscope, with Safety Class II standard, where the oscilloscope will not have its own earth return, and thus the earth of the equipment under test will form the only earth; or you can use an isolation transformer as shown in Fig. 9.5. Either the oscilloscope or the test item can be operated through the isolation transfor-mer, and the other connected normally with its protective earth system. The ground reference connection from the oscilloscope probe will then ensure that both units are at safety earth potential. Figure 9.6 shows the connections for this situation, in this case with the oscilloscope connected through the iso-lation transformer.

APPLICATION EXAMPLE

Diagnosis of faulty power supply – faulty capacitor

The circuit of Fig. 6.5 is again used for this example and is repeated here as Fig. 9.7.

The fault symptoms are similar to those in the application example in Chapter 6 – there being a ripple voltage on the +5 V regulated supply.

The Channel 2 probe is connected to the output terminal of the 7805 regulator, a.c. coupled, and once again there is a ripple wave-form present. The timebase is set to 5 ms/div and the waveform period measured. The period is found to be 2.0 divisions at 5 ms/div, so the period is $2 \times 5\,\text{ms} = 10\,\text{ms}$. This gives a ripple frequency of

$$\frac{1}{10 \times 10^{-3}} = 100\,\text{Hz}$$

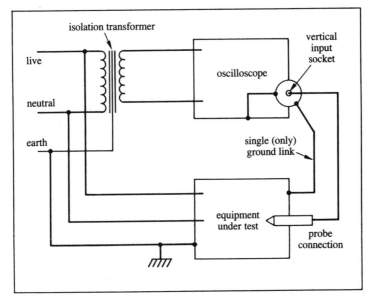

Figure 9.6. Diagram showing the use of a mains isolation transformer for avoidance of ground loops.

The lower trace of Fig. 9.8 shows the waveform at the regulator output. The Channel 1 probe is referenced on the graticule with the input coupling set to ground; then on the d.c. position, the probe is connected to the input terminal of the regulator. The upper trace in Fig. 9.8 shows the input waveform to the regulator.

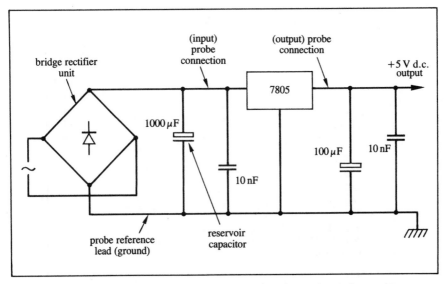

Figure 9.7. Circuit diagram of a 5 v d.c. power supply using a three pin regulator.

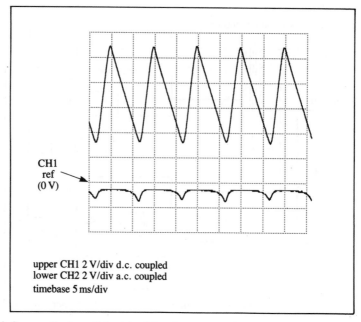

Figure 9.8. Screen display of the 'ripple' voltages on a faulty 5 V d.c. supply.

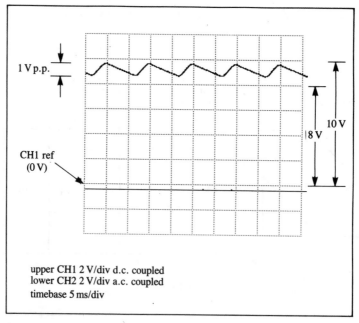

Figure 9.9. Screen display of the 'ripple' voltages on a functional 5 V d.c. supply.

Now the voltage across the reservoir capacitor has a very high ripple level of about 7.5 V p.p., varying from about +3.5 V d.c. to +11 V d.c. The ripple voltage at this point should be about 1 V p.p. under load conditions, and the d.c. level should be at least +8 V.

The 100 Hz ripple frequency does not suggest any problem with the rectifier, and the fact that each successive cycle (at 100 Hz) or half-cycle of mains input has the same amplitude, supports the idea that the bridge rectifier is functioning correctly. The most likely cause of this high ripple voltage is a faulty 1000 μF reservoir capacitor, and the best course is to replace this component.

Once the reservoir capacitor has been replaced, the two probes are reconnected as before. The Channel 2 probe is connected to the regulator output, a.c. coupled, and after checking the ground reference position, the Channel 1 probe is reconnected across the reservoir capacitor. Figure 9.9 shows the two waveforms. The upper trace shows that the output voltage to the regulator now has a +9 V d.c. level, and only about 1 V p.p. ripple level. The lower trace shows that the output is now clean, with no noise or ripple present, so the fault has been cleared.

Either the Channel 1 or Channel 2 probe can now be used to check the d.c. output level by setting the input coupling switch to d.c., and the attenuator to the 1 V/div range. The trace should now deflect upwards (positive) exactly 5.0 divisions from the reference line, indicating a +5.0 V d.c. supply.

10
Storage tubes and sampling oscilloscopes

The fastest growing sector of oscilloscope development is undoubtedly the digital storage oscilloscope (DSO), where new semiconductor technology allows sampling speeds and memory sizes to go higher and higher. Before looking at DSOs, however, we shall consider two other types of oscilloscope which fit between the standard analog oscilloscope and the digital storage oscilloscope.

1 The first of these is the **tube** storage oscilloscope, where the signal was stored within a special storage cathode ray tube itself, rather than in digital semiconductor memories.
2 The second type is the **sampling** oscilloscope, which uses very fast signal sampling and the short-term memory of a charged capacitor to achieve fast effective risetimes.

The word sampling is rather ambiguous in this context since it is used both for the original sampling scopes and the modern digital storage scopes. As we shall see later, in the 'analog' sampling scope, sampling is where a *small sample of one cycle of the input waveform* is taken. Then on a later cycle of the input waveform, another small sample is taken, just a bit later on in the waveform cycle. On a later cycle of the input waveform, another small sample is taken, again just further on in the cycle than the last, and so on. These successive samples are then displayed on the screen as a continuous event, and they reconstruct the original waveshape. The samples are analog levels representing the vertical points on the waveform at particular moments in time.

In the case of the DSO, the sample rate refers to the clock speed at which the analog-to-digital converter operates. The input signal feeds into the a-to-d converter and in this case there are *many samples per one cycle of the input waveform*. These samples are a digital representation of the input signal.

The digital storage scope has now virtually replaced storage tube oscilloscopes, and in some cases has combined with sampling scopes to produce an even better combination, utilizing the benefits of high speed sampling techniques with the ability to digitally store the samples.

Let us first look at tube storage scopes.

10.1   Storage tubes

Before the digital revolution, the limitations of the ordinary analog oscillo-scope had to be overcome by other solutions. Two major shortcomings of the standard oscilloscope are as follows:

1 *Viewing of slow signals*. When slow timebase speeds are used, the display becomes a spot of light moving across the screen instead of a solid trace. As the spot traces out the waveshape, each part fades and disappears immediately, so the eye and brain cannot retain the image of the overall waveshape. This problem can be partly overcome by using a long persis-tence phosphor on the CRT faceplate. Special phosphor compounds are used which glow for longer after being struck by the electron beam. So when the spot produces the illuminated trace on the screen, it will stay for perhaps a few seconds before fading out. Of course, this is only a partial solution because at very slow timebase speeds the spot may only travel a centimetre every few seconds, so the problem remains.
2 *Single events*. A second limitation of the standard oscilloscope is when viewing single or intermittent events. The single shot mode of the timebase will allow a solid triggered display, but for only one sweep of the timebase. Once the event is completed, it is lost. Again, there is a solution to this problem, but rather an inconvenient one: it is to attach a camera to the front of the oscilloscope and record the single event on film as it sweeps across the screen. This involves the cost of the photographic system and film, and the delay while the pictures are processed. The advent of polar-oid film solved the delay problem, but the cost and inconvenience remain.

The storage tube dealt with both of these problems by storing the analog waveform within the tube itself, and provided many other benefits as well. There are two main types of storage tubes used in oscilloscopes: the direct view bistable type and the transfer type. We shall just look at the direct view type as an example of how a tube can be used to store a waveform, and only a simplified description will be given.

The storage tube is rather more complicated and much more expensive than the standard oscilloscope tube, and considerably more circuitry is required to operate it. The main differences are the extra storage mesh and backing elec-trodes behind the faceplate, and the additional flood guns which illuminate the stored picture (see Fig. 10.1).

The main cathode and plate assembly at the rear of the tube is the same as is found in a standard tube, and in this case forms the writing gun. This gun behaves in the normal way with the function of generating, focusing and deflecting a narrow beam of electrons which is then accelerated towards the faceplate. But in this case the (writing) electron beam has an extra function which is to establish a charge pattern on that part of the storage mesh that it

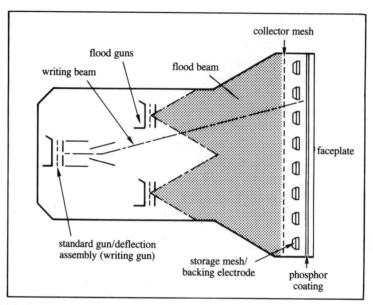

Figure 10.1. Construction of a typical direct view bistable tube.

passes through. The storage mesh is situated just behind the front screen of the tube, and is a very fine wire mesh parallel with the faceplate. On the rear (gun) side of this mesh is deposited a thin layer of very low leakage dielectric material. Slightly behind this storage mesh is another mesh, the collector mesh. This one has no dielectric deposit, but is a plain metal mesh. The other main difference in this tube is the addition of a pair of electron guns situated nearer to the front of the tube than the writing gun. Their function is to produce wide unfocused beams known as the flood beams. The two flood guns are positioned so that their output beams overlap and produce one large flood beam which covers the whole of the screen area. The role of this flood beam is to reproduce the stored image on the screen according to the charge pattern on the storage mesh.

The storage electrode system works rather like the control grid in the ordinary tube. The storage mesh and backing electrode voltages are critical in the control of the writing and flood beams. Assume first that the writing beam traces out the waveform on the tube screen in the normal way. The storage mesh is at a voltage slightly negative with respect to the flood gun cathodes. The collector mesh is positive with respect to the storage mesh and flood gun cathode. The faceplate is at a high positive potential (about 5 kV) and acts as a final anode or accelerator. As the fine electron writing beam passes through the storage mesh, it also collides with the dielectric layer in the areas immediately surrounding the holes in the storage mesh. These tiny areas are then charged positive by the beam, and thus form a positive charge pattern in the shape of the written waveform. Where the waveshape has charged an 'image' as a positive charge pattern on the backing electrode, this pattern will now be positive with respect to the flood gun cathode. Where the positive potential exists on the backing electrode, the electrons will be attracted from the flood gun cathode (backing electrode waveshape is positive, cathode is negative). Where there is no charge on the backing electrode, the electrons will be repelled by it (backing electrode is negative with respect to cathode). So the flood beam will pass through the collector mesh, and where the positive charge exists (the waveshape) be attracted to, and through the storage mesh, to go on and strike the screen phosphor. Where the backing electrode is still negative due to not having the waveshape charged on it, the flood beam electrons will be repelled back towards, and collected by the collector mesh.

Thus, where the positive (waveform) charge exists, the flood beam is transmitted towards the faceplate, and accelerated by its high positive potential to strike the screen. Thus, the previously (once) written waveform is continuously traced out by the flood beam.

By varying the potential on the backing electrode, it is then possible to erase the charge pattern on the backing electrode by restoring the whole dielectric layer to the same potential. A new waveform can then be charged on to it by the writing beam. After capture of a single event waveform, storage times in excess of an hour can be obtained without too much picture degradation.

By using pulse techniques for the erase operation, a variable persistence mode can be achieved, which is useful for the optimum viewing of signals in the range approximately 0.1–50 Hz. The variable persistence mode is where the time that the stored image stays visible can be varied between (say) 1 millisecond and 100 seconds. Then, on which ever timebase range is used, the storage time can be set so that as the spot reaches the right of the screen, the stored image is just starting to fade on the left. Each successive new write cycle of the sweep will then display the new waveform without quite overlapping the old one.

It can be seen that these storage tubes are able to store waveforms for long periods just as the modern digital counterpart does. They also have extra advantages such as variable persistence as mentioned above, signal retention even when unpowered and fast writing speeds. This means that very high frequency signals can be stored. They also have various disadvantages such as high cost, weight, fragility and eventual swamping of the stored picture. This is where the screen background floods up to swamp the stored trace with an evenly illuminated screen.

There was a further development of the storage tube which is less easily replaced by digital technology, and this is the transfer tube. The particular advantage of this transfer tube is its very high writing speed. The tube uses different cycles for its writing and viewing phases. By first writing on to a storage mesh or target, without any screen display, input frequencies in excess of 100 MHz can be stored, and single events only a few nanoseconds wide can be captured. Once the signal is stored, the view mode is automatically cycled, transferring the stored waveform to the screen and allowing long-term examination of the signal.

The second type of oscilloscope we shall look at is the sampling scope which permitted the effective bandwidth of an oscilloscope to be extended to above 1 GHz many years ago.

10.2   Sampling oscilloscopes

Sampling scopes are neither digital nor analog storage scopes, but normal analog oscilloscopes using a sampling technique. Sampling scopes are a means of examining a very fast repetitive signal using an oscilloscope whose analog bandwidth does not extend to those high frequencies. There are various types of sampling scope, but the ones we shall consider are the sequential sampling mode, random sampling mode and real time sampling types. For ease of understanding, the first one we shall look at is the sequential sampling type.

10.2.1 *Sequential sampling mode*

Let us first assume that the state of the art limit of analog bandwidth is 100 MHz. But we want to look at a signal of 1 GHz (1000 MHz). The 1 GHz input signal to the oscilloscope is first fed from the Y input socket to a sampling gate. The sampling gate is rather like an on/off switch connected to a capacitor. When the switch is on, the input signal goes through the switch and connects to the capacitor, thus charging the capacitor to the signal voltage level at that moment. When the switch is turned off, the signal is disconnected from the capacitor, leaving the capacitor charged to the signal voltage level at the precise time of disconnection. The sampling gate is not actually a switch but a diode gate which is switched on by a very fast pulse, and the duration of the pulse is very small compared to the signal period. So the capacitor only charges to a voltage representing a small part of one cycle of the input signal (see Fig. 10.2).

When the sampling command pulse turns off the gate, the capacitor stays charged to the voltage level of the signal sample. Then, after the sampling gate is turned off, the input carries on through its cycle, and in fact many more complete cycles of the input signal occur before another sample is taken.

Meanwhile the sample voltage on the capacitor is transferred through to the vertical amplifier of the oscilloscope and displayed on the screen. The sampling scope timebase performs in a special way in that the sweep wave-

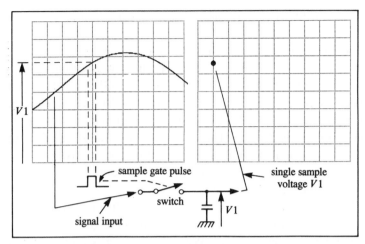

Figure 10.2. Single sample taken from a sinewave signal and held as the charge on a capacitor.

form, instead of being a linear ramp, is a staircase waveform. This is similar to the standard sawtooth ramp but with small steps superimposed on it. When the ramp is rising, the (unseen) spot is moving left to right across the screen, but when the voltage is at a 'stair' level, it is staying at a constant voltage for a short while. So if the voltage is momentarily constant, there is no movement of the spot in the horizontal direction. This is the point at which the signal voltage sample is shown on the screen. When the sample reaches the Y plates of the tube, the spot is unblanked and thus becomes visible. At the same time the horizontal movement momentarily ceases as explained above, so the bright spot appears at a fixed point on the screen.

Now after (say) 10 cycles of the input signal, the trigger point on the eleventh cycle starts the next command pulse sequence in operation to take a second sample. This next sample is taken a little bit later on in the signal period compared to the first sample.

The sample transfer, timebase ramp and step, and unblanking sequence all repeat as before. So one more sample is displayed on the screen slightly to the right of the first. Meanwhile, 10 more signal cycles enter the input socket. As you can imagine, a third sample is taken after 10 more input cycles, slightly later in the input signal period than the second sample; and the unblanking of the spot reveals a third display point on the screen.

The process then repeats over and over until the display consists of a complete waveform of dots. Figure 10.3 shows how the sequential samples from progressively later points on the waveform are displayed to look like a single cycle of that waveform. For the sake of simplicity, only 10 cycles are shown forming the new displayed waveform, but in practice hundreds of samples may be taken and displayed across the screen. Also the samples may be taken from a different multiple of inputs such as every hundredth cycle, etc.

When the display is complete in the above example, let us assume that there is one complete cycle of the input occupying the whole screen width. Now we know that the input signal frequency is actually 1 GHz, so the signal period must be the reciprocal, $1/f$. Thus the period 1×10^{-9} s which is one nanosecond. Since the graticule is 10 divisions wide, and one cycle of the waveform takes exactly one screen width, the effective scan time is equal to the signal period which is one nanosecond. With a scan of one nanosecond for 10 divisions, the effective timebase speed is thus 1 ns divided by 10 which is 100 ps/div (100 picoseconds/div).

Figure 10.3 shows the sequential samples taken from each tenth input cycle to form the dotted waveform. We will assume here that only 10 cycles are taken for the whole timebase scan. When the vertical sample command pulse briefly closes the switch, the signal level at that time is transferred to the capacitor. Then when the switch is disconnected, the voltage on the capacitor is sent at a slower rate to the CRT Y plates and placed on the screen as a dot. Now we are assuming that one sample is taken from every tenth cycle of the signal which is 1 GHz. The input signal period is therefore 1 ns. Therefore, a sample is taken once every 10 ns, and the total time to take the 10 samples is 100 ns. In real time, the timebase takes 100 ns to place the 10 dots across the screen, but as we saw above the effective timebase speed is 100 ps/div.

We assumed above that the oscilloscope has a bandwidth of 100 MHz, so its risetime would be 350/100 (ns) = 3.5 ns. So as the sweep moves across the screen, there is 10 ns between each sample for the spot to move from one position to the next, and the amplifier risetime is only 3.5 ns. So this technique has allowed us to view a signal much faster than the vertical amplifier bandwidth would normally reach. Although only 10 samples were shown in the example, in practice up to 1000 dots may be displayed to form a continuous trace.

Having developed the sampling technique for the evaluation of very fast signals, a further refinement was added which gave the benefit of pretrigger display and this was the random mode sampling technique.

10.2.2 *Random mode sampling system*

The random mode sampling system is *not* a method of displaying random input signals. The input signal has to be repetitive just as with the sequential mode system. Also the manner in which samples are taken is not random, the samples being obtained in a very precise controlled way. When the samples are displayed on the screen, they do not appear in the strict left to right sequence as before, but seem to appear at any position in the waveform. This apparent random order of appearance is what gives the name to the system

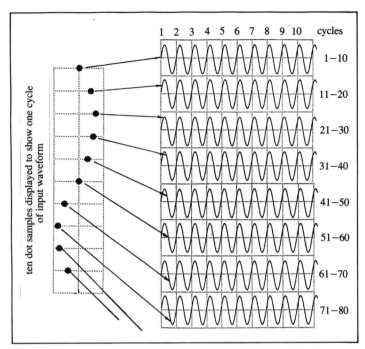

Figure 10.3. Sequential sampling of a signal over many cycles to form a displayed single cycle.

'random mode'. Look at Fig. 10.4. The input signal is used to generate a precise trigger point in the normal way, and this trigger point initiates a timing ramp. However this timing ramp is not fed directly to the horizontal deflection plates. Instead, the timing ramp voltage level is sampled at a fixed time after the vertical sample.

So with the random mode sampling system, *pairs of samples* are taken, one of the input waveform voltage level and one of the timing ramp voltage level, although these two samples are not taken simultaneously. The two sampled voltages are displayed simultaneously as the y and x coordinates, the tube being momentarily unblanked to display the spot.

As in the sequential system, when the trace is unblanked, both the y and x coordinates are held constant, so each displayed spot does not move while it is visible. The y coordinate is a fixed voltage sample from the input signal and the x coordinate is a fixed timing ramp sample voltage. The input signal must be periodic, and as such has a fixed time between given points on consecutive cycles of the waveform. This time from the trigger point (say) to the same point on the next cycle of the signal is its period. Now if a time delay is introduced that is just less than the signal period, the end of this delay will occur just ahead of the next trigger point.

This principle is used to generate a pretrigger system so that samples of the waveform can be taken before the trigger point. In fact the time delay is an exact multiple of the signal period (less a little bit) so that samples are taken every 10 cycles (say) just as with the sequential sampling system. So the end of this time delay occurs just before the trigger point 10 cycles later.

Starting from the point where the time delay ends, let us consider the sequence of events beginning with samples that are taken before the trigger point. Due to this delay system the sample timing control comes from the trigger point on a previous cycle, perhaps many signal cycles before, so we will have to assume our start after a trigger point, many signal cycles and the delay period.

First, the end of the time delay executes a vertical sample command pulse, so a vertical sample is taken and this voltage is stored as the charge on a capacitor. This is our first y coordinate. This point also starts a time sequence t_s, whose end executes the horizontal sample command.

Next the input signal crosses the trigger threshold and generates a trigger pulse which initiates the timing ramp.

When the time t_s ends, the horizontal command pulse occurs and a sample is taken of the timing ramp voltage at that time. This voltage is again stored as the charge on a capacitor and forms our first x coordinate.

The CRT is then briefly unblanked, and a spot displayed on the screen at a position determined by the y and x coordinates above. Then many signal cycles later, another vertical signal sample is taken, but now the delay time is varied slightly so that this sample occurs at a different point on the input waveform and it is not a repeat of the first sample. Again, after time delay t_s, another timing ramp sample is taken and stored, and later this pair of y and x coordinates position the spot on the tube face while it is unblanked.

The vertical sample is taken with reference from and thus later than the trigger point. But since it occurs several cycles later due to the delay timer, it can now occur before the (later) trigger point. If the delay time is increased very slightly with each sample that is taken, eventually the samples will occur later and later in the signal cycle, from before, to during and after the trigger point. When these samples are all displayed on the screen with their associated timing ramp sample voltages, a complete waveform display is formed showing the waveshape before, at and after the trigger point.

As with the sequential sampling system, the effective display sweep rate is very fast. But now with random sampling method the signal can be seen before and after the trigger point as well.

Random sampling scopes are equipped with a Time Position control to adjust the proportion of pretrigger display on the screen. This is the control that adjusts the time delay between the trigger point and the vertical sample on a later cycle. There is also the Time/Div control which like the sequential system, sets the *effective* scan rate across the screen.

Figure 10.4 shows two examples of vertical samples taken in the random

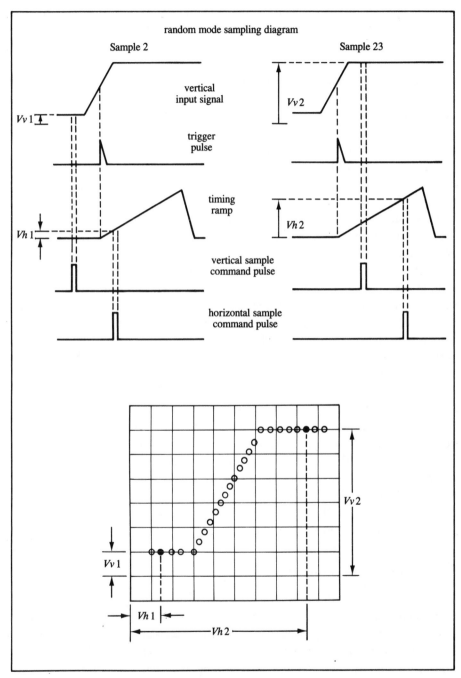

Figure 10.4. Timing relationships in random mode sampling, with resultant screen display.

sampling mode. These two samples are *not* in sequence, but demonstrate one sample taken before and one after the trigger point. In the waveform display at the bottom, the two samples are the second and twenty-third of the twenty-five displayed, and are shown solid.

Vertical sample 2 is taken before the trigger point *on that signal cycle*. The vertical sample command pulse is timed from the trigger point on a previous cycle. When the vertical sample command occurs, the input signal voltage is sampled as $Vv1$ and stored. There is then a time delay determined by the front panel controls, and after this time, the horizontal command pulse occurs. The voltage level of the timing ramp is sampled as $Vh1$ and stored. These two voltages $Vv1$ and $Vh1$ are then applied to the vertical and horizontal deflection plates (after amplification) and the trace unblanked. Sample 2 is thus displayed on the screen.

Now many more samples occur in the same way, each vertical sample being delayed slightly more with respect to the trigger point, so that each sample occurs later and later in the signal cycle.

Vertical sample 23 is a case where the delay has increased and the vertical sample $Vv2$ is taken well after the trigger point. So when the horizontal sample command occurs after the same time period as before, the timing ramp, initiated at the trigger point, has reached a much higher voltage level. This timing ramp sample $Vh2$, when applied to the CRT with the associated vertical sample $Vv2$, thus positions the spot much further to the right. So when the spot is briefly unblanked, coordinates $Vv2$ and $Vh2$ place the spot at the positions shown solid as sample 23.

Thus the Time Position control allows the trigger point to be horizontally positioned on the screen, and the sampling system allows very fast signals to be examined. This combination makes the random mode sampling system an outstanding method for the examination of 'state of the art' signals.

Having developed sampling scopes for the display of very fast signals, their applications were limited to just that. In order to extend their versatility, another operating mode was incorporated to allow use at lower frequencies. This is the real time sampling mode.

10.2.3   Real time sampling

When a sampling oscilloscope is used for its primary purpose, high frequency input signal samples are only taken once per many cycles. At lower signal frequencies this sampling procedure is neither necessary nor desirable. So the sampling system can be switched to operate in a different way known as real time sampling. In this mode, the sampling frequency is controlled by an oscillator which runs at a high frequency compared to the input signal. It is very similar to the chop oscillator system in normal vertical amplifier channel switches. The sampling oscillator may be free running or synchronous with the sweep cycle. The timebase is running and displaying in real time, not in effective time as before. The real time sampling scope behaves just as a normal analog scope, with a triggered sweep system and switched timebase speeds. Now with the real time sampling system, the vertical signal is sampled many times per cycle and with the free running oscillator type, the samples are taken and displayed in the same X position and with the same scan speed. In fact it behaves just as a normal analog scope except that instead of the Y input signal passing straight through to the Y plates, it is sampled and every sample is passed through to the Y plates. Since there are so many samples taken per cycle of the input signal, they join together to form a continuous real time waveform. There is no synchronous relationship between the sample rate and the input, trigger or sweep signals.

With the synchronous real time sampling method, the sampling pulses are locked to the sweep cycle. So as the sweep occurs, each sample represents a discrete time segment relative to the sweep start point. If these sampling pulses are counted for the duration of one sweep, a direct digital code is provided which exactly indicates the sweep rate in digital format. These data can then be used for a digital readout display of the sweep rate.

APPLICATION EXAMPLE

Testing frequency response of an audio amplifier

In this example we shall check the frequency response of an audio amplifier. Apart from the oscilloscope and the audio amplifier itself, you will need a sinewave generator with a frequency range of at least 1 Hz to 50 kHz and an 8 ohm high power resistor. This resistor is used as a dummy load across the output terminals in place of the speaker.

The signal generator is used to supply the input signal instead of the normal audio signal from the compact disc or auxiliary source. The signal generator must be set to a (low) level within the acceptable range of the amplifier you are testing so that it does not exceed the maximum input level and cause distortion. A high power 8 ohm resistor is connected across the speaker terminals of one channel. The resistor should be rated at least 1 watt and preferably higher. It should be securely connected across the output terminals free from contact with anything else as it may get very hot.

Before starting the measurement, all the tone and equalizer controls must be set to their mid or off positions so that the main amplifier response is tested without the effect of these controls. The typical input voltage level required by an amplifier into its compact disc or aux inputs is about 500 mV, so first set the output level of the sine-wave generator to approximately 500 mV r.m.s. If there is no output level indication on the generator, first feed the signal directly into one of the oscilloscope inputs and, with the attenuator set to the 500 mV position, adjust the generator for about two divisions amplitude peak to peak. This will give a level of 1 V p.p. which is about 350 mV r.m.s. and within our 500 mV requirement. Then remove the signal from the oscilloscope. The sinewave frequency should then be set to 1 kHz, and the output signal now connected into one of the amplifier inputs, let us say the left channel. It may be necessary to make up a special connector lead to make the link between the signal generator and the left channel compact disc or aux input sockets.

Next fit the 8 ohm resistor to the respective output, in this case the left channel. The oscilloscopes should then be connected as shown in Fig. 10.5. The Channel 1 probe is connected across the signal gener-ator terminals to monitor the input signal, and the Channel 2 probe across the resistor to monitor the amplifier output.

Set the Channel 1 attenuator and variable amplitude controls to obtain a trace of exactly 2.0 divisions. Keep the timebase to a fairly low speed, say 10 ms/div, so that the display appears as a band of signal, rather than individual cycles. This makes it easier to set the

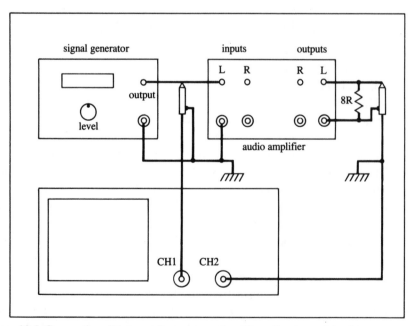

Figure 10.5. Connection diagram of apparatus for the evaluation of the frequency response of an audio amplifier.

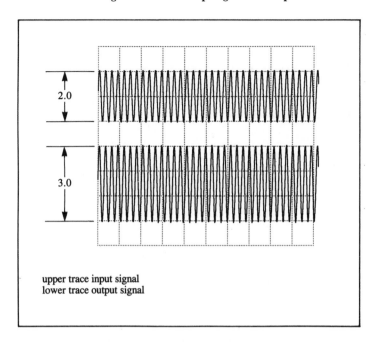

Figure 10.6. Screen display of the input and output waveforms of an audio amplifier at a frequency of 1 kHz.

amplitude exactly to the 2.0 divisions. Once this setting is achieved, adjust the Channel 2 attenuator and variable controls so that the output signal is displayed at an amplitude exactly 3.0 divisions peak to peak. Figure 10.6 shows the two traces at this point. The upper trace, Channel 1, shows the input signal as a band at 2.0 divisions amplitude, and the lower trace, Channel 2, the output at 3.0 divisions.

Now leave the oscilloscope controls at these settings during the complete measurement.

First, to find the low frequency −3 dB limit of the amplifier, adjust the signal generator frequency slowly downwards, while watching the amplifier output on the Channel 2 trace on the scope. The signal generator output level must be kept constant so keep watching the Channel 1 trace as well while the frequency is being lowered. If the Channel 1 display amplitude varies from 2.0 divisions at any time, restore it by adjustment of the generator output control. *Do not alter the oscilloscope controls.*

Keeping the Channel 1 trace thus maintained (if necessary) at 2.0 divisions, gradually reduce the signal frequency until the Channel 2 display (the output) falls to exactly 2.1 divisions. Make sure Channel 1 is still at 2.0 divisions. Now this is the −3 dB low frequency limit and should be noted. If the signal frequency is not indicated on the

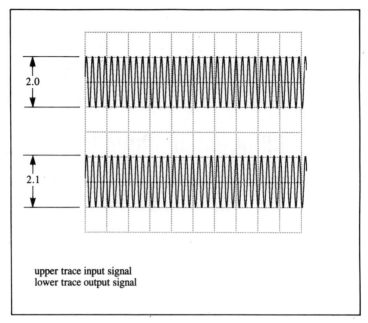

Figure 10.7. Screen display of an audio amplifier input and output waveforms at the $-3\,\mathrm{dB}$ points.

generator, it can be determined by measuring the signal period (one complete cycle) using the timebase, and taking the reciprocal to give the frequency.

Suppose one signal occupies four divisions at 20 ms/div, the period is thus $4 \times 20\,\mathrm{ms} = 80\,\mathrm{ms}$. Therefore

$$f = \frac{1}{t} = \frac{1}{80 \times 10^{-3}} = 12.5\ \mathrm{Hz}$$

The next step is to find the upper frequency limit, so now increase the signal frequency above the audio band again keeping the input level (Channel 1 trace) constant at 2.0 divisions. Set the timebase switch to $100\,\mu\mathrm{s}$/div so that a band of signal is displayed as before. Slowly increase the generator frequency until the Channel 2 trace falls to 2.1 divisions as before. Make sure Channel 1 is still at 2.0 divisions. Now when the output (Channel 2) has fallen to exactly 2.1 divisions, measure the signal frequency. Again, if it is not directly shown, use the oscilloscope to measure it. Suppose that the signal period is now 3 divisions at $10\,\mu\mathrm{s}$/div, the period is thus $3 \times 10 \times 10^{-6} = 30 \times 10^{-6}\,\mathrm{s}$. Therefore

$$f = \frac{1}{30 \times 10^{-6}} = 33.3\ \mathrm{kHz}$$

Figure 10.7 shows the two traces at the frequency limit point, with the upper trace, Channel 1 is still at 2.0 divisions, and in the lower trace, Channel 2 is reduced to 2.1 divisions.

The whole procedure can be repeated for the right channel of the amplifier by transferring the input signal and probe to the right input, and the 8 ohm resistor and probe to the right speaker output terminals. Unless there is a fault condition, the response of each channel of the amplifier should be almost identical.

11
Digital storage oscilloscopes

The current range of digital storage oscilloscopes that are fitted with a normal vacuum CRT (as opposed to LCD displays) usually perform as normal analog scopes as well, with many of the normal functions already described in previous chapters. They also have the ability to sample the analog signal and convert the samples to digital form. These samples can then be stored in a random access memory (RAM) in the form of a single integrated circuit (IC) or 'chip'. The data stored in the RAM can then be read out, converted back to an analog signal, and fed to the oscilloscope screen display. By using several of these RAMs, many different waveforms can be simultaneously stored in memory, and retrieved when required.

Although there are many benefits to the digital storage oscilloscope, there are two main advantages over analog scopes. These are:

1 The ability to observe slow and very slow signals as a solid presentation on the screen. Slow moving signals in the 10–100 Hz range are difficult to see and measure on a normal analog oscilloscope due to the flicker of the trace and the short persistence of the spot on the screen. Very slow moving signals, less than 10 Hz, are impossible to view on an analog scope. As fast as the spot traces out the waveform, the image fades and disappears before a complete picture can be formed. Both these types of signal can be viewed on the storage scope just as solidly as a 1 kHz sinewave.
2 The ability to hold or retain a signal in memory for long periods. This can range from fractions of a second to hours or longer, depending on how the memory supply voltage is maintained. With battery back-up, this can extend to weeks or months. Also the digitized signal can often be fed out of the oscilloscope for retention or manipulation elsewhere, for example to computer disk, magnetic tape, hard copy printer, and so on.

The key things that determine the suitability of a storage oscilloscope to a particular application are the sampling rate and memory size. The memory size refers to the amount of horizontal segments of the trace that can be divided into one sweep of the timebase. For example if the memory size was 1k (1000), then the trace would be divided into 1000 horizontal segments.

With a standard 10 cm trace, this means $1000/10 = 100$ samples per centimetre. With 100 samples taken in 1 cm, they are probably just close enough together to look like a solid trace. Now if the memory was doubled to 2k, then there would be 200 samples per centimetre. So the definition would be twice as good.

The sample rate is usually determined by a crystal controlled clock oscillator. The crystal ensures accuracy, making sure that the master clock is always the same fixed frequency, although the clock signal fed to the digital processing system may be a sub-multiple of the oscillator frequency according to the timebase speed.

Now assume that the clock frequency is 1 MHz, and the memory size is 1k (1000). With a clock rate of 1 MHz, each sample occurs in one-millionth of a second, i.e. $1 \mu s$. To fill up a memory of 1k (1000 samples) takes $1000 \times 1 \mu s = 1 ms$. So one whole sweep of the timebase would fill the memory in 1 ms. Now there are usually 10 divisions across the screen, so the sweep speed is $1 ms/10 = 100 \mu s/div$ (maximum). Slower sweep speeds can then be selected, simultaneously dividing the clock rate by the same factor as the time/div switch. For example on $200 \mu s/div$ range, the clock would be halved to 500 kHz to fill the 1k memory in one sweep of the timebase. Slower and slower speeds can be used indefinitely, by dividing the clock sampling rate, but the maximum sweep speed is determined by the maximum clock speed as shown above.

Assuming a memory size of 1k, we are thus taking 1000 samples of the input waveform as the spot makes one sweep across the screen, left to right. For each one of these sample points, a digital code is stored, representing the input voltage. Now the vertical screen display is usually 8 divisions of 1 cm each. The most common method is to use an 8 bit digital signal to encode the analog signal, so with an 8 bit binary code representing the vertical position of the spot, the resolution can be determined. In binary code, 8 bit is 2 to the power of 8, i.e.

$$2^8 = 256$$

So the vertical display is divided into 256 vertical points. Hence for an 8 cm or 80 mm screen height, and allowing another 1 cm overscan above and below to give 10 cm or 100 mm display height, the resolution of the vertical display would be:

$$100/256 = 0.4 \text{ mm}$$

better than 0.5 per cent accuracy.

For each one of the 1000 samples taken horizontally across the screen, an 8 bit digital code is stored in memory, representing one of the 256 vertical positions on the screen. So the memory is known as $1k \times 8$, and thus holds

8000 'bits' of information. (In fact digital memories are binary multiples, e.g. $1024 \times 8 = 2^{10} \times 2^3$, so a $1k \times 8$ memory is really 1024×8.)

During the transition of the spot across the screen of the CRT, the up and down movements are changed to a digital coded signal by an analog-to-digital converter. This is fed to the memory, and by the end of the sweep the memory is full. If the memory contents are now not locked in by operating the 'hold' or 'save' control, the next sweep of the trace will overwrite the memory with the new (or replaced) signal on the screen. Very slow moving signals are fed into memory as they happen, but once stored they can be displayed on the screen at a much faster rate. So very slow signals can be observed as easily as fast signals.

The sample rate of DSOs currently available varies from 10 MHz at the low cost end, up to 1 GHz (one thousand million samples per second). The maximum sample rate quoted in specifications is sometimes for single channel performance. In dual trace mode, the sampling rate may be halved, as two memories are filled on alternate cycles of the clock. Other manufacturers maintain the same sample rate whether one, two or more channels are in use. In this case, the input signals are simultaneously sampled on each cycle of the clock.

The maximum input signal frequency that can be stored is not a fixed figure, or ratio of the sampling clock speed. In general it is advisable to keep the input signal frequency to the oscilloscope at least 10 times smaller than the sampling rate, and if possible at least 20 times smaller. If the maximum sampling rate of a DSO is 20 MHz, then this means that the maximum stored signal frequency should be 20 MHz/10 = 2 MHz (1 MHz preferred). This is often known as the maximum effective storage frequency and would thus be calculated as the maximum sampling rate divided by 10. Thus in the example above, the maximum effective storage frequency = max sampling rate/ 10 = 20 MHz/10 = 2 MHz. In practice, it is possible to use higher input frequencies if the signal shape is known. A sinewave will allow storage up to about a quarter clock frequency (in this case 5 MHz) if the storage timebase ranges go high enough. By using the dot join facility, a waveform with relatively few samples can be joined up (linear interpolation), and if the signal is a sinewave, the nature of the signal is known. In that case a better technique is sometimes available to enable accurate interpolation, or 'fill in' between the dots. This is using the function $\sin x/x$ or curve interpolation, where each successive coordinate of the waveshape is a function of the last. Using this technique, higher frequency sinewaves can be displayed for a given sampling rate, while still producing a satisfactory waveform on the screen. Since the sinewave shape is known, and there are no higher frequencies contained in the signal, the few samples taken in each cycle of the waveform are sufficient to give an indication of the waveshape and size. It can be shown that 3.5 samples per cycle are required, so in this case the maximum effective storage frequency

is given by maximum sampling rate/3.5. So in this case using curve interpolation, the maximum effective storage frequency = 20 MHz/3.5 = 5.7 MHz.
However, the highest timebase range on the instrument is usually limited in
the storage mode to fill the memory size in one sweep of the timebase at
maximum sample rate. So when assessing the high frequency and glitch
capture capability of a DSO, there are three factors to consider. These are:
maximum sampling rate; memory size; and fastest storage mode timebase
speed.

Let us take an example. Suppose the storage specification for an instrument
is as follows:

$$\text{Maximum sample rate } S_r = 20 \text{ MHz}$$
$$\text{Fastest storage } T/B \text{ speed } TB = 10 \,\mu\text{s/div}$$
$$\text{Memory size } M = 2\text{k} \times 8$$

The overall sweep time (T_s) for a 10 division horizontal sweep will be
$TB \times 10$, here $10 \times 10^{-6} \times 10, = 100 \times 10^{-6}$ (100 μs). If we now multiply the
sweep time T_s by the sample rate of the clock SR, we can find the number of
samples N in one sweep, i.e.

$$N = S_r \times T_s$$
$$N = 20 \times 10^6 \times 100 \times 10^{-6}$$
$$N = 2000$$

So the number of samples taken is 2000, and hence the 2k memory is filled in
one sweep. Now if we apply the (storage) maximum input frequency factor of
10 as indicated above, we divide the clock rate by 10.

$$20 \text{ MHz}/10 = 2 \text{ MHz}$$

On this basis, the maximum frequency input signal we can reasonably store is
a 2 MHz sinewave. This 2 MHz signal will have a period of

$$1/f = 1/2 \times 10^6 = 500 \times 10^{-9} \text{ (500 nanoseconds)}$$

The maximum storage timebase speed above was 10 μs/div, so in one division
we will display

$$10 \,\mu\text{s}/500 \text{ ns} = 10 \times 10^{-6}/500 \times 10^{-9} = 20 \text{ cycles}$$

With a display of 20 cycles of input waveform per division (200 across the
whole screen) each individual cycle is difficult to see and resolve. There may
be a ×10 horizontal magnifier on the oscilloscope, so using this to increase the
scan to 1 μs/div, there are two cycles per division displayed. Now with one
complete cycle of the input waveform only half a division wide on the screen,

with only 20 samples in it, there is little that can be learnt from the display apart from its frequency and size.

If the input waveform is not sinusoidal, it may be impossible to determine the exact nature of the signal, and the low sample density may mean that some waveform details are 'missed' by the samples. So for non-sinusoidal signals, keep the input frequency at least 20 or more times lower than the sample rate.

11.1   Equivalent sampling

You may have read in the last chapter about the analog sampling scope system and its advantage of increasing the effective bandwidth. The same system is used in digital storage scopes to achieve the same results. Instead of each successive sample being fed into the digital memory from the a-to-d converter, many cycles of input signal are allowed to 'go by' before the next sample is stored. In this way, many thousands of input signal cycles may occur before the memory is filled. For instance the a-to-d converter may sample a point on cycle '1' of the input signal. The second sample point, slightly later in the input signal cycle, instead of being taken on cycle '1', is delayed until cycle '101'. Then by sampling every hundredth input cycle and storing it, the memory is gradually filled. Using this equivalent sampling mode, the maximum frequency input signal that can (theoretically) be stored is increased by 100 (in this case). In practice the samples may be taken every tenth cycle, or hundredth, thousandth, or a binary multiple such as 256, 1024, etc.

In this example, using equivalent sampling, effective storage frequency = max equivalent sampling rate/10. Using the previous example of our 20 MHz DSO, max equivalent sampling rate = 20 MHz × 100 = 2 GHz. Then the effective storage frequency = 2 GHz/10 = 200 MHz. So by using equivalent sampling, the highest input frequency that can be stored is raised to 200 MHz. In practice, however, the upper frequency limit will usually be determined by the analog bandwidth. The analog bandwidth of the scope used in this example may be 100 MHz, 50 MHz or only 20 MHz the same as the digital sampling rate, so this is the figure that must be used as the limiting frequency.

11.2   Glitch capture

A similar exercise is involved in the determination of the glitch capture capability of the digital storage oscilloscope. A glitch is a non-repetitive, irregular, spike type pulse that may occur in a circuit. It is unwanted but is found sometimes in digital systems, for instance when clock edge transitions occur. It is necessary to detect these glitches and 'design them out' of the system before the circuit can be considered reliable.

Returning to the maximum sample rate, taking the frequency this time to be 40 MHz, the period is thus $1/f$

$$= 1/40 \times 10^6$$

thus period

$$= 25 \times 10^{-9}$$
$$= 25 \, \text{ns}$$

So the narrowest glitch that could be recorded is 25 ns wide. In practice a better rule of thumb for glitch capture, is to double this figure, and say minimum glitch capture is twice the sample period. In this case, a 40 MHz clock gives (2 × 25 ns) 50 ns glitch capture. These figures are not absolute, and must be applied with discretion to the circumstances involved. It rather depends upon the shape of the glitch, the risetime, falltime and width (if any) as to whether a 25 ns or greater glitch can be recorded with a 25 ns sample.

A glitch that was repetitive and occurred every 25 ns could be considered to be the limit for a 40 MHz sample rate, but if there was only one (dot) sample per glitch, nothing would be known about it except its presence. To gain some detail about the shape and amplitude of the glitch, we would need perhaps a minimum of four samples. So with our 40 MHz sampling rate, a sample is taken every 25 ns, so four samples take 100 ns. You might consider, therefore, that for glitch capture with waveshape interpretation, a 40 MHz sample rate enables 100 ns glitch capture.

In practice this may not be the only constraint. With a 2k memory size, and a 25 ns sample period, the memory will be loaded in 2000 × 25 ns = 50 μs. With a 10 division sweep, this is equivalent to 5 μs/div.

To put these figures in perspective, an input waveform which was displayed as one cycle per division at the fastest (digital) timebase speed of 5 μs/div would thus have a period of 5 μs.

Therefore, the frequency would be

$$\frac{1}{t} = \frac{1}{5 \times 10^{-6}} = 200 \, \text{kHz}$$

A glitch of 25 ns displayed on this speed range would then be very narrow. With a time range of 5 μs/div, a 25 ns pulse would only be

$$\frac{25 \times 10^{-9}}{5 \times 10^{-6}} = 0.005 \, \text{divisions wide}$$

If the oscilloscope was fitted with a ×10 horizontal magnifier, this could be used to increase the pulse display to 0.05 div (0.5 mm) wide. So even now, with the oscilloscope used to its full capacity, this fast glitch is displayed only 0.05 mm wide, so very little can be learnt from it apart from its presence.

The problem is that glitches, by their nature, are non-repetitive. If they were recurring signals, there would be no need to use digital storage at all. A fast analog display would show all the detail required about a glitch. However, since they are such intermittent events, it is necessary to use a combination of single sweep and digital storage to capture and hold these glitches for examination. It is important to consider all the above factors when determining the minimum specification for glitch capture. It should be clear from the examples above that the glitch capture capability should be considered as a combination of:

1   Digital sample rate.
2   Fastest (digital) timebase speed.
3   Horizontal memory size.
4   (Expected) glitch width.
5   (Expected) glitch repetition rate.
6   X Amplifier magnification.

There is another factor which can prove vital in this glitch capture process – electronic data manipulation. After having stored a random signal within the constraints outlined above (1 to 6), it is possible with some systems to magnify the (horizontal) signal electronically. This is done by reading the samples out from the memory at a slower rate compared to the horizontal scan rate. So the waveform is 'stretched out' on the screen. This facility is commonly available as a further magnification of 10 times on the horizontal display, and may be carried out within the oscilloscope, or using software in other equipment such as a computer or printer.

It should be remembered that when data are output from the oscilloscope to other equipment, say a PC, the horizontal magnifier effect is not transferred with the data. In the oscilloscope, the magnifier comes after the signal is converted back to analog, and so this magnifier is operative on the oscilloscope display only. Thus a 10-fold magnification electronically by a PC will be the same as the ×10 magnifier on the oscilloscope.

When the *oscilloscope* has the software magnification built in, however, giving a ×10 expansion (say), a further ×10 magnification can also be obtained from the oscilloscope's normal (× amp) magnifier control. So in this case, with software expansion *and* horizontal magnifier expansion, a factor of ×100 expansion may be possible after glitch capture.

In the example above, with a 40 MHz sample rate, 2k memory and 5 µs/div fastest speed, this means the fastest display rate would then be 5 µs/div times 100 = 50 ns/div. In order for glitch capture to be successful on a digital storage oscilloscope, it usually requires the use of the single sweep storage, and hold functions, so we shall go on to see how these special facilities operate.

Having looked at the background to digital storage oscilloscopes, let us now consider some of the controls associated with their operation.

11.3   Store

This is the main control for switching from analog-to-digital operation. Although the oscilloscope is operated in the same general manner in both modes, care must be taken as some controls may have a different function or effect in digital mode to that in analog mode. For example, the timebase variable control will work normally in the analog mode, while having no effect at all in the storage mode. This is because the analog sweep generator is completely disabled in the storage mode, and the sweep is formed from the sampling clock generator, with fixed frequencies set by the time/division switch. Hence, in storage operation, the variable is out of circuit. Other functions may also be affected, such as the trigger level, single shot and so on. All the time the instrument is in the storage mode, the display is the memory content being continuously re-displayed, not the direct vertical input signal.

The most common storage mode is known as REFRESH, where the sequence of events is as follows: the input signal fires the trigger circuit and the subsequent trigger pulse initiates the memory write cycle. The analog-to-digital converter changes the input signal to the 8 bit code which is fed into the memory until it is full. Once the memory is full, the write cycle stops, the digital code is read back out, converted back to analog form and displayed on the screen.

Now on the receipt of the next trigger signal, the memory is refilled with the next signal, or refreshed. If the trigger input stops, the write cycle stops, but the read cycle continues. Now the d-to-a converter keeps re-reading the memory contents and displaying the same signal on the screen. As long as the input and consequent trigger signal continues, the refresh cycle continues, each sweep displaying the new memory contents.

11.4   Hold or save

Most digital storage scopes operate on an automatic 'refresh' system. That is, every new sweep of the timebase refills and overwrites the store with the new input signal. So if you wish to retain a particular signal in the store you must stop it being overwritten. The HOLD or SAVE button achieves this by 'locking' the store so no new signal can get in. This control can be used in conjunction with the trigger level and the single shot functions to lock the store at the right moment between access of the required signal, and the overwriting by a later signal.

A similar effect is achieved by removing the input connection. Without the trigger pulse to start a new write cycle, the memory contents do not refresh. However, this is not a safe condition, as any following spurious noise, pick-up or alteration of the trigger controls may cause a trigger pulse which would start a write cycle. The stored signal is then lost. So always use the hold or save control, to be sure to lock in the memory contents for as long as you need to.

11.5   Dot join

Since the signal displayed on the screen in the storage mode is recovered from the digital memory, it must therefore consist of a string of samples of the analog waveform. However, between these samples are gaps in the waveform (where no sample was taken). The dot join facility puts a line between the samples to join up the dots into a continuous trace (linear interpolation). This has the advantage of making the display brighter and easier to interpret. However, it should be noted that the dot join function only 'fills in' the gaps between samples with a bright line, which does not represent what the signal actually did in that small period of time. The dot join facility is electronically generated in the oscilloscope, so the 'filled in' lines between samples of the signal are just straight line joins between the sample points, and not related to the input waveshape (see Figs 11.1 and 11.2).

Figure 11.1 shows a stored waveform with no dot join. Although the gaps are exaggerated for effect, it is clearly a sinewave. Figure 11.2 shows the same waveform with dot join turned on, and the straight line joins are thus shown up. In practice the effect is negligible since the dots or samples are so close together. In the diagrams only about 20 samples are shown across the whole screen to emphasize the effect, but in practice there are at least 1000 samples. On some oscilloscopes there may be up to 8000 samples in one horizontal sweep to fill an 8k memory, and in future these memories will surely get larger.

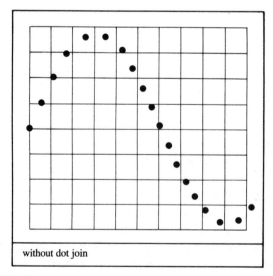

without dot join

Figure 11.1. A digitally sampled waveform displayed without dot join.

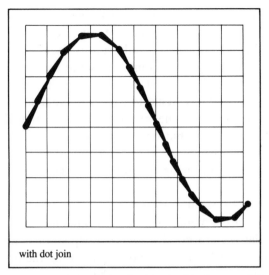

with dot join

Figure 11.2. A digitally sampled waveform displayed with the use of dot join.

When the input signal is at maximum practical frequency, that is about one-tenth sample rate, the dot join must be used carefully. Under these conditions, the number of samples per cycle of input waveform is small. So without dot join in use, the few dots represent points on the input waveform. But straight line joins between the dots may give a completely false impression of the waveshape, particularly if the waveshape is unknown and irregular. So exercise caution when interpreting waveforms under these conditions.

If the waveform is known to be a sinewave, there is another method of 'filling in' the gaps sometimes available. This is 'sine' join, where the mathematical function $\sin x/x$ is used to determine the shape of the curve between the samples (curve interpolation). It is a feature found on some storage oscilloscopes, but it is only appropriate for sinewave inputs. Since the shape of a sinewave is precisely known, each gap in the sampled waveform can be filled with a curved line to match the characteristic sine shape.

11.6 Single sweep

The single sweep function may occur on different oscilloscopes as a function exclusive to the storage mode, exclusive to the analog mode, or applicable to both. It is with respect to the storage mode that we are concerned here.

Firstly, the single sweep control must be operated in conjunction with the trigger level. The 'single' facility always has the associated 'RESET' function with it. Once 'single' is selected, the system must be 'armed' to wait for the desired signal. To allow this, the timebase must be taken from the automatic trigger mode (where the timebase would run by itself and record no input) to the manual mode. The trigger level must be set to the optimum point to allow the desired signal to 'fire' the timebase. The single mode is then selected and the reset depressed. The next signal to trigger the timebase is displayed on the screen and also loaded into the memory. The signal will now stay in the memory as long as the reset button is not depressed again. Selecting the save or hold for that memory allows the stored signal to be retained for longer still, and prevents it from being overwritten, at least until the instrument is switched off.

A particular benefit of single sweep storage is to avoid double triggering of the sweep. In earlier chapters, the use of variable hold-off was described, for use with particular waveforms. The single sweep storage mode can be used to achieve the same result by triggering on, and storing the first group of pulses, then 'locking out' subsequent pulses. The signal in the memory is then continuously displayed on the screen.

There are some special functions which are only possible on digital storage oscilloscopes. As already described, the storage facility can simply be used to store or retain a signal displayed on the screen, that is, the same triggered signal can be kept, and later re-displayed in the same manner. However, for slow moving signals, another display mode is often used, called the ROLL mode.

11.7　Roll mode

When the roll mode is used, the input signal is not triggered at all. Instead, a display is provided which enables the changing input to be observed as it slowly changes. This 'real time' type display of the changing signal occurs at the right-hand edge of the screen. It is stored as it happens, then the stored signal is 'rolled' across the screen right to left. This roll effect across the screen gives a particularly informative view of the behaviour of a very slow moving signal.

The effect is rather like watching a pen recorder. Imagine the pen positioned at the right-hand edge of the screen, and the paper moving right to left across the screen, showing the recorded signal being traced out by the pen on the right. Looking at the left shows what happened before, watching the right shows what is happening at the time.

The hold or save control can be utilized when required, to 'freeze' the screen display and subsequently process the stored signal as usual.

On some digital storage oscilloscopes, it is possible to use single sweep in conjunction with the roll mode. When this is done, depressing the reset button 'rolls' one screenful of signal across the screen, right to left, then stops until the next reset command.

The roll mode is only effective on very slow signals, usually less than about 10 Hz. If faster signals are observed in the roll mode, the effect is just like an untriggered display on any analog oscilloscope, it being impossible to interpret the signal.

11.8   Pretrigger

Another feature peculiar to digital storage is the pretrigger function. The name pretrigger refers to the fact that the events that occurred pretrigger, that is, before the trigger point (in the normal triggered mode) can be displayed. The pretrigger selection is usually made as a percentage selection. The value selected is that percentage or proportion of one sweep duration, for whatever speed is used. So on 1 second per division timebase speed, the total sweep time would be $10 \,(\text{divisions}) \times 1 \,(\text{s/div}) = 10 \,\text{seconds}$. So if the 50 per cent pretrigger was selected, then 50 per cent of 10 seconds would be displayed = 5 seconds of signal event *before the trigger point*.

Since the display time always starts on the left, the trigger point effectively moves across the screen as more pretrigger is selected. For example, if the pretrigger is set to 25 per cent, then the trigger point will be 2.5 divisions across from the left-hand edge of the screen. If 50 per cent is chosen, the trigger point will be at the horizontal screen centre, and so on. The pretrigger range is 0–100 per cent, usually in 10 or 25 per cent steps.

Figure 11.3 shows an intermittent signal captured in the single shot storage mode with normal trigger (no pretrigger). When the signal occurs, it only triggers the timebase when its amplitude is sufficiently large to exceed the trigger threshold level. So when the timebase starts, the signal may have already been active, but at an amplitude lower than the trigger threshold set on the oscilloscope.

Figure 11.4 shows the same event captured using 50 per cent pretrigger. The point on the waveform that triggered the sweep in Fig. 11.3 can now be seen at the horizontal screen centre. You can also see the behaviour of the signal prior to the trigger point, displayed to the left of the screen centre. In the example it can be seen that there is a small positive peak, followed by a large negative peak, prior to the trigger point. This part of the signal was not visible in Fig. 11.3 with the trigger point at the left-hand edge of the screen.

In some cases it may be necessary to use (say) 75 per cent pretrigger, to devote 75 per cent of the screen width to the activity before the trigger point, and if necessary to use a slower timebase speed to capture the whole event.

The pretrigger selector can be used in conjunction with the single shot mode, so that even for non-recurrent events, as well as capturing a one-off event in store for subsequent evaluation, the signal activity just prior to that event (at the trigger point) is also stored.

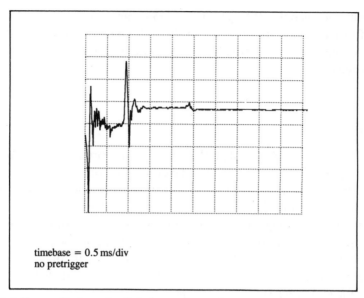

timebase = 0.5 ms/div
no pretrigger

Figure 11.3. Screen display of a digital stored waveform, captured without PRETRIG-GER.

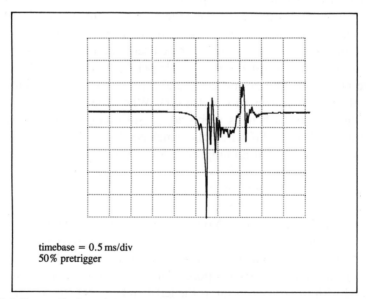

timebase = 0.5 ms/div
50% pretrigger

Figure 11.4. Screen display of the same waveform as in Fig. 11.3, captured with 50 percent PRETRIGGER.

11.9   Aliasing

There is an undesirable effect that can occur in the digital storage mode which is not found in normal analog operation known as aliasing. This is the display of an apparent signal which does not actually exist. The problem is usually caused by wrong setting of the timebase speed range. In order to ensure accurate representation of an analog signal in a digital memory, remember that many samples should be taken per cycle of the input waveform. If, however, the sample rate is much too low, so that only one sample is taken per cycle, or one sample per several cycles, then aliasing occurs. Clearly, if only one cycle is taken per three cycles (say) of input waveform, there is no information obtained on the behaviour of the waveform during these three cycles. But worse still, the samples taken every three cycles may form together, particularly when using dot join, to look like a valid waveform. So care should be taken to ensure that the displayed signal is the right one. Fortunately it is quite simple to check whether the display shows a real signal or an 'alias'.

The quickest way to check for alias signals is to switch back to analog mode and observe the display *on the same timebase range*. The display frequency should be approximately the same in storage and analog modes. (There may be a small difference due to the use of the variable time control which is only effective in analog mode.) If there is a vast difference between the analog display and the store display, then the store signal is probably an alias.

Another way to check is by utilizing the fact that only a real signal can trigger the oscilloscope. So, if in the storage mode, the trigger functions are altered, the display will be affected. For instance, switching between internal and external trigger source will cause loss of trigger in the latter case. If the store signal is unaffected by operating the trigger functions, then it is an alias. (Ignore the trigger indicator lamp as it will be activated by the real input signal, even though you may be seeing an alias.)

Figure 11.5 shows an apparent sinewave signal in the store mode. There are approximately 22 cycles displayed and the timebase speed is 10 ms/div. So the period of the waveform is the total sweep time divided by 22 (cycles), thus the period must be

$$\frac{100 \times 10^{-3}}{22} = 4.54 \text{ ms}$$

So the frequency of this waveform *appears* to be (from $f = 1/t$)

$$f = \frac{1}{4.54 \times 10^{-3}} = 220 \text{ Hz}$$

Now if the oscilloscope is switched from store mode to analog, with the timebase kept at the same speed of 10 ms/div, the waveform then appears as a solid band of signal, where the cycles are so close together as to be undiscernible as in Fig. 11.6. The signal frequency is obviously much higher than it seemed in the store mode. If the timebase speed is now increased step by step until the range of 10 μs/div is reached, the display appears as in Fig. 11.5. Now if the store

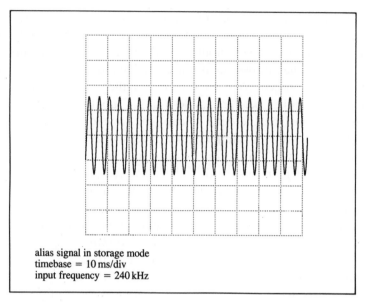

alias signal in storage mode
timebase = 10 ms/div
input frequency = 240 kHz

Figure 11.5. Apparent (ALIAS) signal displayed in STORE mode at 10 ms/div timebase speed.

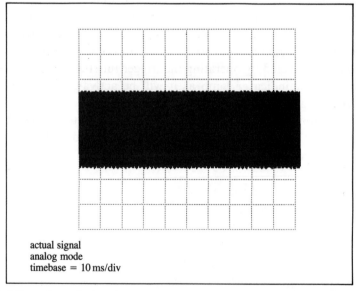

actual signal
analog mode
timebase = 10 ms/div

Figure 11.6. Same (actual) signal as shown in Fig. 11.5, displayed in NON-STORE mode at 10 ms/div timebase speed.

mode is selected, the display remains the same. Figure 11.7 shows the display with the timebase at 10 μs/div, in store mode or analog mode.

So the alias signal appeared at a timebase speed of 10 ms/div, and the real signal was displayed at 10 μs/div, 1000 times faster. Although the displays of

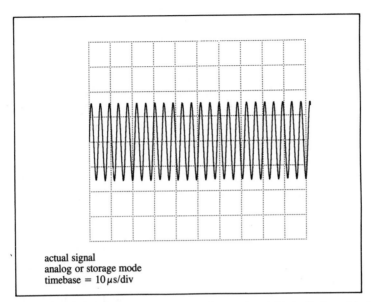

actual signal
analog or storage mode
timebase = 10 μs/div

Figure 11.7. Same (actual) signal as shown in Figs 11.5 and 11.6, displayed in STORE mode at 10 μs/div timebase speed.

Figs. 11.5 and 11.7 look very similar, only the last one is the true waveform. So although the waveform in Fig. 11.5 looks real enough, it does not represent the input signal to the oscilloscope at all. It is worth periodically checking when viewing new signals in the store mode to ensure that you are seeing the real signal. As stated above, just quickly switching from store mode to analog will normally verify the display as genuine, then switch back and continue your measurement.

11.10 Averaging

When a signal is repetitive but 'noisy', it is possible to extract the signal from random noise using an averaging technique found on some oscilloscopes. This averaging function exploits the fact that the noise superimposed on a signal is random, and that over a period of time, the average of a random noise signal is zero. That is, if you look at a specific small part of a signal, over and over again as the signal repeats, the noise signal on that small part will be sometimes positive and sometimes negative. If you look at this small section enough times, all the positive noise signals added together will approximately equal the sum of the negative noise signals at this point. Therefore, the average of these positive and negative noise signals at this point will be virtually zero.

A sample of the waveform is first taken and stored in the memory. Then a second sample is taken at exactly the same trigger point as the first. The two samples should be identical in terms of signal content, but not in terms of noise content. Now the second sample is added to the first, and the result divided by two. Since the two signals are the same, adding and dividing by two (averaging) gets back to the original waveform. So in terms of the signal waveform itself, there is no alteration at all. However, the noise on the signal will be modified. In fact over only two samples, the average noise voltage at a given point on the waveform may be worse than the first sample alone. If we now take two more samples and then average over the whole four, once again the actual signal waveform remains unchanged. The four identical signal voltages added and divided by four yield the same unchanged signal. However, now the four noise signals will start to average out to zero. At a given point on the waveform, the noise voltage superimposed may be equally positive or negative, so after many successive cycles of the waveform, the noise voltage at that given point will have about as many positive as negative excursions. So the more samples that are taken, the better the positive and negative noise signals will cancel out.

Oscilloscopes fitted with this averaging function will typically average over a high binary number of samples such as 256 (2^8). Averaging is not something that can be achieved on all storage oscilloscopes, but is peculiar to those that feature Average or Averaging as a specific front panel function.

The examples given in this chapter referred to an 8 bit code to represent the vertical position of the waveform at any given moment. An 8 bit binary code gives 256 levels, so the screen height is divided into 256 segments. With a vertical scan of (say) 100 mm, this gives a vertical resolution of $100/256 = 0.4$ mm. Thus vertical changes in the signal smaller than 0.4 mm cannot be resolved. However, oscilloscopes can (and do) use higher codes such as 10, 12, 16 and 32 bit, to obtain better resolution and accuracy. As computer technology produces faster and higher resolution digital systems, so the DSO will follow.

APPLICATION EXAMPLE

Evaluating the upper frequency limitations of your DSO

In this example we will evaluate the upper frequency limitations of a digital storage oscilloscope in the store mode using a function generator. Unlike the analog oscilloscope, digital storage capabilities are dependent upon the user's requirements and acceptance level. For a given sampling rate, the upper frequency limits and pulse storage capability are somewhat dependent on what information the operator needs from the display. With a repetitive signal, the maximum frequency that can be stored depends upon the signal shape and the rise and fall times. So we shall test the DSO with various inputs to learn its limitations for future reference.

The DSO we shall test in this example has a 20 MHz sampling rate and a $2k \times 8$ bit memory per channel. It does not matter what the specification is for the model you use, but if it has a higher sampling rate, you may need higher frequency test equipment than stated below. We shall test the oscilloscope first in the single trace mode, and then in dual trace mode to see if there is a difference.

To test the instrument you will need a function generator with a maximum frequency of about 10 MHz and sine and square wave outputs, or a sinewave generator up to 10 MHz and a square wave generator up to about 2 MHz.

First set the oscilloscope to the following conditions:

1   Single trace (CH1) only.
2   Auto trigger (CH1).
3   Store mode.
4   Fastest (digital) timebase speed.
5   Dot join off.

Set all the other controls to the normal, CALibrated or off positions.

Now in this example the fastest digital speed is $10 \, \mu s/div$, so with the timebase switch set to this range, connect the function or sinewave generator to the Channel 1 input socket, a.c. coupled. Adjust the generator output amplitude and the Channel 1 attenuator to give a display amplitude of 5 divisions peak to peak. Use the Channel 1 position control to set the display at the screen centre. Now adjust the generator frequency to about 100 kHz and check the oscilloscope display. You should now see about one cycle per division across the screen. Select the $\times$ magnifier and increase the generator frequency by the same factor. In this example the $\times$ magnifier is $\times 10$, so the sinewave frequency was increased by a factor of 10 to 1 MHz. Once again the display shows about one cycle per division.

The oscilloscope is now operating at its fastest storage capability, so now gradually increase the generator frequency and observe the effect on the display. The horizontal density of the signal will increase to the point where separate cycles are not discernible. Also the vertical amplitude will begin to decrease as the signal frequency increases. When the display reaches the point you consider to be the limit of usable information, stop and note this frequency.

Having reached this point, now switch back to the analog mode, and observe the effect on the display. Note that any change to the display is due to the limiting effects of the digital storage system. Having noted this difference you may wish to reconsider your noted upper frequency limit for sinewave storage, and perhaps make it a little lower, so switch back to store mode and set the input to a frequency you know will be realistic when compared in the analog mode. In the example this frequency limit was found to be approximately 2 MHz.

Now change the input signal to a 100 kHz square wave either by selecting square function on the generator, or using a separate generator if necessary. Leaving the oscilloscope settings as they were, slowly increase the square wave frequency until again the display is only just an acceptable representation of the square wave signal. Again switching over to analog mode will check what the signal should be like, and if necessary revise the frequency setting in store mode. Now note this frequency as the top limit for storage of square or pulse type signals.

You might find it beneficial to keep holding or saving the test waveform each time as this will 'freeze' the samples on the screen. This may give a worse display and cause you to go lower in frequency, but it will be a more practical result as it may be an intermittent pulse that you need to store in future, and the saved display may be the only choice.

In the example, the top acceptable frequency was found to be about 1.0 MHz.

Once you have determined your two frequency limits for the instrument you can apply them with discretion to each application you have in future. You need to remember why you chose the frequencies as the limiting points. If it was because the amplitude was falling, and gave unrealistic readings, you may choose to ignore this factor and exceed the frequency, but keep in mind the effect. If the signal was lacking definition due to insufficient sample content, again you can ignore this when you know the shape of the signal, and indeed use dot join to improve the display.

However, beware of exceeding the frequency limits and using dot join when the signal shape is not precisely known. Often you will be using store mode to save a signal *because* its shape is unknown, so in this case interpret the display with care.

If the sampling rate is clearly specified as the same in single or dual trace mode, then the results that you have obtained will be the same. However, if the sampling rate is halved in dual trace mode or if you are not sure, then it is best to check the performance in the dual trace mode to see if the limits are different. So if this is the case, proceed to check on dual trace mode.

Return all the instrument settings to the same conditions 1 to 5 above. Now also select dual mode, and use the Channel 2 position control to set the second trace on the bottom graticule line, out of the way. Then repeat the same tests as before on the Channel 1 trace. If you obtain different results this time, then the sampling rate is probably halved in the dual mode. There is no need to check Channel 2 as it will be identical to Channel 1 in dual mode. When you reach the limiting frequencies in each case in dual mode, check the difference by momentarily switching back to single trace (CH1) then back to dual.

Now you may have two sets of figures to remember for future applications, and again these can be used to advantage as you will always be able to get optimum performance from the instrument. For instance, if you are working on an application with two traces displayed, you can briefly switch to single mode to enhance the display, and then switch back.

12
Choosing an oscilloscope

If you are about to choose an oscilloscope to buy, there are several basic factors to be considered in order to get the right one. The first priority is to buy the best you can afford. Although a complex oscilloscope may be more difficult to operate at first, and have special facilities not needed at present, everything will be there if you need it later. Better than than having to keep changing up in specification because the scope cannot do the job you need it for. So get the highest specification model you can afford in the first place, from a reputable manufacturer with proven reliability.

If you need an oscilloscope for one specific task or group of tasks only, then it will be most economic to buy a model just able to deal with those requirements. Also, if the instrument is to be used by untrained personnel, where the controls need to be as simple and few as possible, the minimum basic specification is required. In these two cases it will be necessary to choose between a basic model for the sake of cost or simplicity, and a more advanced instrument. The restricted specifications of a basic oscilloscope may make it unsuitable if the requirements change, so a compromise may be preferable, perhaps choosing a model a little more advanced than the current requirements demand.

There are hundreds of oscilloscopes to choose from, with a very wide range of specifications, so you need to sort out your own specification requirements. If you can afford a top range model that does everything, all well and good, but if not you must make some choices.

12.1 Analog or digital

If you choose a digital storage scope, then it will almost certainly also be an analog scope too, and probably with a superior analog-to-digital performance in terms of its frequency range. So the choice is really between an analog scope and a digital/analog combined model. So do you need digital storage? Some definite reasons for choosing digital storage are as follows:

1 *Viewing of low frequency signals.* Viewing signals lower than about 50 Hz is much easier with digital storage, as a permanent display is presented, rather than a moving spot.

2 *Permanent record of waveforms.* If you want to keep a record of your waveform displays, digital systems allow short-term memory storage in semiconductor memories, and long-term in computer files (disks) and printers.

3 *Data manipulation.* Digital data information, equivalent to an analog display signal, can be output to computers for mathematical processing and analysis.

4 *Recording of single event phenomena.* Intermittently occurring signals can be captured from just one occurrence of the event, then continuously viewed.

If you have no, or very little need for any of the above applications, due to the higher cost of digital storage scopes, choose an analog model!

12.2 Single, dual or multitrace

A single trace oscilloscope does have a few advantages:

1 *Simplicity of operation.* Because of the minimum number of controls and facilities, they are very easy to operate.

2 *Inexpensive.* The cheapest models are usually the single trace units, although a little more money buys many more advantages.

3 *Lightweight.* If a really lightweight portable is needed, it is often found as a single trace unit.

Obviously the great disadvantage is the single trace itself, which does not allow comparison of waveforms. Often for very little extra expense, a dual trace model can be found.

Dual trace units are by far the most popular general-purpose oscilloscopes, combining the advantages of dual displays without the complexity of extra traces and extra functions which may not be required.

The great advantage of a dual display is the comparison of one waveform with another. A signal can be compared before and after processing in some way, or for timing reference along the waveform. Often one channel is used as a reference, displaying a particular waveform, while the other channel is connected to various different parts of a circuit.

One channel can be used as a trigger source channel, while the other is used for signal examination. Although the EXTERNAL TRIGGER facility can be used for this purpose, waveforms are much more comprehensible if you can see the triggering reference waveform. The fact that the dual trace oscilloscope has become the industry standard is evidence enough of the virtues of this type of instrument.

For some applications, two traces may not be enough, and there are a variety of multitrace oscilloscopes available to fill this need. For digital and logic applications, four, six, eight and more traces are available. These oscilloscopes are ideal for logic analysis, timing comparison and so on, and consequently have many inputs. However, these inputs may be logic (TTL) level only without attenuators, and therefore not suitable for general use. Multitrace instruments are obviously very useful for special applications, but because of the many extra front panel controls required, they are too complex for normal general use. So unless more than two traces (or possibly three) are actually required, choose a general-purpose, dual trace oscilloscope.

12.3   Mains or battery portable

Only a very small proportion of oscilloscopes on the market are battery operated portables. There are some which are both mains and battery, but since they suffer the constraints of the battery portables, they should be considered as such.

The one (and only) advantage of the battery powered scope is that you can operate it anywhere, away from mains sources. This is essential if you are working in a remote location or in a safety area (where high voltages are not permitted). But a high price must be paid for this capability, notably screen size and performance. The main problem with making a battery powered portable oscilloscope is the power requirement of the CRT. In order to get a reasonable working period from a set (or charge cycle) of batteries, the power consumption must be kept as low as possible; and this means relatively slow speeds and performance. This would also be at a higher relative cost than the mains powered counterpart.

The problem is being overcome by the use of LCD (liquid crystal display) screens, since they consume much less power than a CRT. Models are now available with sample rates of 50 MHz and more, giving effective usable bandwidths of up to 20 MHz. However, these LCD portables are currently above five times the cost of an equivalent bandwidth mains powered analog instrument. So unless mains independent portability is absolutely required, stick to a mains powered model for the best price/performance deal.

12.4   Bandwidth (and risetime)

The most important specification by which oscilloscopes are categorized (and costed) is the bandwidth. This is the upper frequency limitation of the instrument. As a general rule, get the highest bandwidth you can afford. The input waveform is unaffected by the timebase speed or the trigger, but it is greatly affected if its frequency approaches the oscilloscope bandwidth. If the input signal is a pure sinewave, then its amplitude will simply reduce as the bandwidth limit is approached. However, for any other shaped waveform, the displayed shape can be totally distorted by the frequency limitation of the oscilloscope's vertical amplifier system. So choose an oscilloscope with a bandwidth well above the top frequency you want to work at.

The risetime of the vertical amplifier is a direct function of the bandwidth, so if you have chosen your required bandwidth, the risetime is fixed for you (see Chapter 8, Sect 8.2.1 and Table 12.1).

12.5   Sensitivity (vertical deflection factor)

Unless you want to look at very small (or very large) signals, choosing sensitivity is not a problem as most standard oscilloscopes are very similar. The most common (maximum) sensitivity is 5 mV/div with possible amplification to 1 mV/div by a magnifier system. Since it is recommended always to use a ×10 probe, the maximum sensitivity with the probe in use would thus be reduced from 5 mV to 50 mV/div. If you check signals smaller than 50 mV peak to peak, then you need an instrument with higher nominal sensitivity, say 1 mV or 2 mV per division; or a 5 mV/div model with a magnifier. If you use mainly low frequency signals in the audio range, it may be possible to use a ×1 probe, but beware of the bandwidth limitations and loading effects of these probes (see Chapter 3 on probes). If you mainly use very large signals, the minimum sensitivity will be of more interest, and you should look at the most counterclockwise position of the attenuator. With a typical minimum range of 5 V/div, and using a ×10 standard probe, this gives 50 V/div. With 8 divisions of vertical display, you can thus display 400 V peak to peak. Of course ×100 probes are readily available, but then the maximum input is likely to be determined by the voltage limit of the probe, say 1000 V, rather than the maximum display (here 8 divisions at 5 V/div and ×100 probe = 4000 V p.p.). Whenever there are voltage breakdown limitations in a system, the lowest one must be observed.

12.6   Timebase range

If the bandwidth has already been decided, the choice of timebase ranges may be almost nil. Most manufacturers aim to display one cycle per division at the rated bandwidth, so for a 20 MHz oscilloscope, 50 ns/div would be common.

The period of a 20 MHz waveform is 1/frequency. Hence

$$\text{Period} = 1/20 \times 10^6$$
$$= 5 \times 10^{-9}$$
$$= 50 \text{ ns}$$

So to display one cycle at the rated bandwidth, a timebase speed of 50 ns/div is required. This will often be achieved by a top timebase speed of 500 ns/div, and a ×10 horizontal magnifier to increase to 50 ns/div (see Table 12.1). However, faster top speeds may be at the expense of fewer slow speeds on the switch. If you often need slow sweep speeds, make sure that this is a priority when you choose.

12.7   Accuracy

The standard accuracy on most oscilloscopes is now 3 per cent. Under some conditions this may be degraded to 5 per cent, when using magnifier controls for example. Since the oscilloscope is a measuring instrument, the accuracy is vital, so make sure that you do not accept worse than the standard 3 per cent. Read through the specifications carefully and make sure that the maximum accuracy is maintained during all operating conditions of the instrument.

There are now models available with accuracies of 2 per cent and even 1 per cent, but these are very expensive. If this degree of accuracy is required for frequency measurement, it may be better to buy a frequency counter and a cheaper 3 per cent accuracy oscilloscope. Even very inexpensive frequency counters feature high accuracy compared to an oscilloscope, so a combination of the two can be a cost-effective solution.

If high accuracy is required for amplitude measurements, this is not so easily achieved, due to the variable nature of waveshapes. In the case of sinewave measurements, a voltmeter is an accurate, cost-effective solution. However, for other waveshapes, a high accuracy oscilloscope may be the only answer, particularly for selective amplitude measurement and over a wide frequency range.

12.8   Signal delay

Signal delay in an oscilloscope is achieved by the use of a delay line in the vertical amplifier system. This enables the whole of the leading edge of the pulse that triggers the timebase to be displayed on the screen. This is a most useful feature, and is recommended as a high priority when choosing an oscilloscope. Since it is a feature that has no associated controls, is in circuit all the time, and does not make the oscilloscope more difficult to operate, it is well worth having. Most higher performance oscilloscopes now contain a delay line as standard, and some low cost models also have them fitted. If you look at the diagrams in Fig. 2.12 you can see the difference in displaying a pulse with and without a delay line.

12.9   Timebase delay

The choice of an instrument fitted with a delay timebase system is really a major decision since there is usually a large price difference associated with it. If you are looking at the range of oscilloscopes that include timebase delay, then select one that has it. It is one of the most useful functions on an oscilloscope for the isolation and observation of specific details of a waveform. It enables the user to select a small part anywhere along a displayed waveform and expand it across the screen for detailed observation and measurement.

There is the slight disadvantage with a delay timebase that it involves several extra front panel controls which may make the oscilloscope more complicated to operate. However, this is outweighed by the benefits the delay timebase brings to the instrument, so it is really a cost choice only.

12.10 Trigger facilities

Although all oscilloscopes have trigger systems, since they are a key factor for successful operation, it is essential to get an oscilloscope with good triggering. The extra facilities you need will depend on your application of the instrument, so if you are involved with television signals for instance, a TV sync separator is essential on the model you choose.

Trigger filter switches with a.c., d.c., LF and HF positions, etc., enable the trigger circuit to discriminate between different parts of complex waveforms. Often you cannot trigger at all on some signals without using the trigger filter system to reject some part of the waveform, while allowing the signal of interest to get through to the trigger circuit.

A useful point to check is whether on a dual trace instrument, you can trigger from one channel while displaying only the other one. For instance, display Channel 1 only on the screen while triggered on Channel 2. This is a useful facility when the screen is cluttered with overlapping signals, since you can view separately the signals displayed on each channel without having to change the trigger source. (Select Channel 1 only; then select Channel 2 only; then go back to dual.)

If your applications are straightforward, it may be better to keep the trigger extras to a minimum to avoid making the oscilloscope unnecessarily complicated to operate.

12.11 Extra functions

As well as all the features outlined above, there are always extra functions which may be peculiar to each manufacturer's range or model. If you have a special application which is particularly catered for by one of these functions, then it may be a deciding factor in your choice. Each of them can only be judged on individual merit and should only be given attention after consideration of all the main features outlined above.

12.12 Selection table

Table 12.1 shows the relationship between the vertical amplifier bandwidth and risetime, and the associated timebase range required to display one cycle of input signal at the same frequency as the bandwidth over a minimum of one horizontal division. For instance, if the oscilloscope has a bandwidth of 20 MHz, its risetime will be 17.5 ns. To display a 20 MHz signal with a minimum of one division per cycle, a timebase speed of 50 ns/div is required.

Table 12.1. Bandwidth selection table

Vertical amplifier Bandwidth (MHz)	Vertical amplifier risetime (ns)	Timebase speed to display one cycle at max. BW (ns)
5	70	200
10	35	100
20	17.5	50
30	11.6	20
40	8.7	20
50	7.0	20
60	5.8	10
70	5.0	10
80	4.4	10
90	3.9	10
100	3.5	10

In this case a 20 MHz signal at 50 ns/div will give exactly one division per cycle (which in this case is the same as one cycle per division).

If we take a scope bandwidth of 30 MHz, with a risetime of 11.6 ns, the timebase speed for a minimum of one cycle per division is 20 ns/div. The period of a 30 MHz signal is 33.3 ns, so the signal will be displayed with one cycle every 1.65 divisions. Table 12.1 gives the timebase ranges for a display of a bandwidth frequency signal with one cycle at least one division wide. You might find that the top speed to achieve this result is with the horizontal magnifier included. For instance, with a 20 MHz instrument, the fastest timebase range may be 500 ns/div, and a ×10 magnifier is then used to increase the top range to 50 ns/div.

If you are about to buy an oscilloscope, before making your choice you could tabulate your requirements in *your priority order* (see Table 12.2). Then from a shortlist of (say) four preferred oscilloscopes, you can fill in the table with ticks (3) according to the specifications of each model. The scope with the most ticks should then be the best choice. Here is a list of possible factors that you might use as your choices:

- Single trace
- Low cost
- High sensitivity (1–2 mV/div)
- Average sensitivity (5 mV/div)
- High bandwidth (over 50 MHz)
- Low bandwidth (up to 20 MHz)
- Reliability
- Dual trace
- Portable
- Delay line
- Delay timebase
- Basic digital storage
- Fast digital storage
- Extra functions

Now suppose for my future applications I need a dual trace scope with high bandwidth, delay line and delay timebase.

I would like, although they are not essential, low cost, reliability and higher sensitivity. Other facilities would be a bonus.

So now I must list the items with dual trace at the top, single trace at the bottom, high bandwidth second, then delay line and delay timebase. Then I must complete Table 12.2 in priority order, using the specifications of four manufacturers' oscilloscopes which I had shortlisted as being suitable for my needs, calling them scopes A, B, C and D. Now looking at the table, scope B is the best choice having most ticks at the top of the column.

Table 12.2. Factors to consider when choosing an oscilloscope

	A	B	C	D
Dual trace	3	3	3	3
High bandwidth	3	3	3	3
Delay line	3	3		3
Delay timebase	3	3	3	
Low cost		3		3
Reliability	3	3		3
High sensitivity			3	3
Extra functions		3		
Portable				
Basic digital store				
Fast digital store				
Average sensitivity	3	3		
Low bandwidth				
Single trace				

Make your own list of choice factors, arrange them in priority order and thus form your own selection table.

APPLICATION EXAMPLE

Investigation of capacitor–resistor (CR) network using only the oscilloscope internal calibrator as signal generator

If your oscilloscope has a built-in calibrator or probe test point, it will most likely be a low voltage square wave signal at a frequency of about 1 kHz. We will use this as the test signal for our CR circuit. If your oscilloscope does not have this calibrator signal, then you should use a separate square wave generator with an output amplitude of 100–500 mV at a frequency of 1 kHz. The components we will use are a capacitor of 10 nF (0.01 μF) and a resistor of 10 kohms. The **time constant** of this combination is $C \cdot R$ or the product of the two values, so here

$$Cr = 10\,\text{nF} \times 10\,\text{kohms}$$
$$= 10 \times 10^{-9} \times 10 \times 10^{3}$$
$$= 100 \times 10^{-6}$$
$$= 100\,\mu s$$

Looking at our calibrator signal, the period for a 1 kHz square wave is the reciprocal of the frequency, $1/f$, so the period

$$t = 1/1 \times 10^{-3}$$

Therefore

$$t = 1\,\text{ms} \ (\text{or } 1000\,\mu s)$$

So the duration of one half-cycle is half the period, which is 1000 μs divided by two, which is 500 μs.

Now when a capacitor is charged by a voltage, it reaches 63 per cent of the total voltage level in time CR (here 100 μs, see above).

It will charge to almost the full voltage (98 per cent) in five times CR, so in this example with a 100 μs time constant, $5 \times CR = 500\,\mu s$, and this is the same as the time for one half-cycle of our input calibrator waveform.

We can connect the two components in two ways, either with the capacitor in series and the resistor to ground (differentiator), or with the resistor in series and the capacitor to ground (integrator). Figure 12.1 shows the two configurations. The left-hand side is the input and is connected to the square wave signal. The right-hand side is the output and shows the modified waveshape. Both the input (upper trace) and output (lower trace) waveforms are connected to the oscilloscope for observation. As always make sure that your probes are carefully compensated using the probe trimmer adjustment before starting these tests (see Chapter 3 on probe adjustments).

First connect the two components as a differentiator. The square wave signal is connected to the 10 nF capacitor, and the other side of the capacitor is connected both to the oscilloscope input socket (lower trace) and via the 10 kohm resistor to ground. The resultant differentiated square wave is shown in Fig. 12.2. It can be seen that when the fast edges of the square wave occur, the voltage is transferred through the capacitor producing a fast edge at the output. At this point the capacitor has the same voltage level on each side

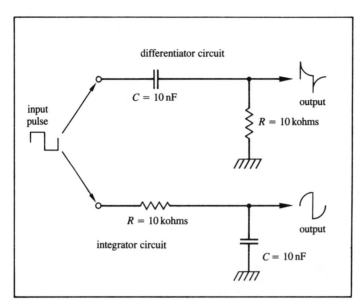

Figure 12.1. Basic differentiator and integrator circuits.

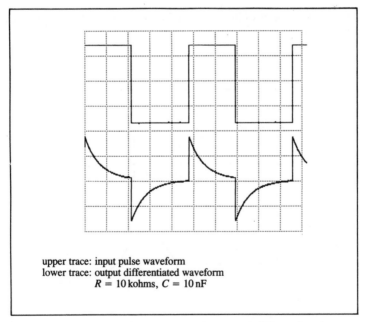

upper trace: input pulse waveform
lower trace: output differentiated waveform
$R = 10\,\text{kohms}$, $C = 10\,\text{nF}$

Figure 12.2. Screen waveforms of square wave input and differentiated output from CR circuit.

(uncharged). Then the capacitor is charged via the 10 kohm resistor with the characteristic exponential shape until the output side reaches ground potential. If the resistor value were to be reduced, the capacitor would charge faster giving a 'spike' type output signal, and this

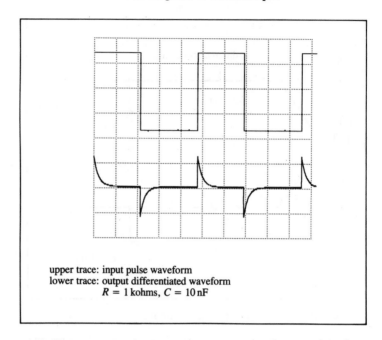

upper trace: input pulse waveform
lower trace: output differentiated waveform
$R = 1 \text{ kohms}, C = 10 \text{ nF}$

Figure 12.3. Screen waveforms of input and differentiated output from CR circuit with reduced time constant.

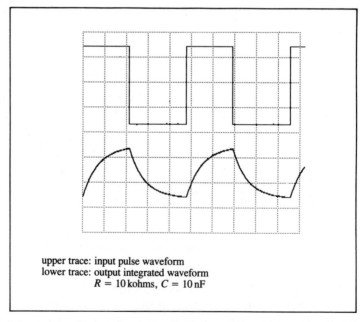

upper trace: input pulse waveform
lower trace: output integrated waveform
$R = 10 \text{ kohms}, C = 10 \text{ nF}$

Figure 12.4. Screen waveforms of square wave input and integrated output from CR circuit.

technique is often used to change a square wave or pulse signal to a spike or needle type signal such as for producing marker pulses. Figure 12.3 shows the effect of reducing the resistor value to 1 kohm, giving the narrow spiky appearance to the signal as mentioned above. If the resistor value were to be reduced further, the needle pulse would become even narrower as the capacitor is able to charge faster, however the amplitude of the pulses will also reduce.

If you compare the upper and lower trace waveforms in each case, you can see the time relationship between the input square wave and the output needle waveform. If the resistor value was increased to say 100 kohms, the capacitor would not be charged in half the input signal period, so the output level will not have reached zero potential when the next fast edge appears and reverses the polarity of the capacitor.

Having seen the effect of a differentiator circuit on a square wave, we will now look at an integrator circuit. Using the original values of 10 kohms and 10 nF, connect the two components as an integrator as shown in Fig. 12.1. The input pulse now connects to the resistor with the output connected through the capacitor to ground. Connect the input and output points to the upper and lower oscilloscope inputs as before and observe the effect of the integrator on the square wave signal.

In this case the current flow into the capacitor is limited by the 10 kohm resistor, so the voltage rises exponentially giving the rounded appearance shown in the lower trace in Fig. 12.4. The capacitor is just about fully charged in half the input signal period as before.

In this case, with the bottom of the capacitor connected to ground, it is uncharged when the top of the capacitor is also at zero (ground) potential. When the input signal changes to a different voltage level, the capacitor then charges towards that level. The capacitor charging current is fed through the 10 kohm resistor charging the capacitor as before in time constant CR.

Again you can experiment by changing the resistor value and noting the effects on the input waveform.

13
Calibration – how to check it

For an oscilloscope to be an accurate measuring instrument, and not just a display system, it is essential that it is calibrated within the manufacturer's specified limits. Once an instrument is more than a year old, or if you have any reason to suspect the calibration accuracy, it is worth checking the main parameters to make sure that all is well. It can be very costly to have to repeat measurements taken over a long period, which are later found to have been made with an inaccurate instrument.

The three main parameters which need to be checked to ensure measuring accuracy are: horizontal time measurements; vertical voltage measurements; and vertical bandwidth. The last two factors are related.

The vertical voltage measuring accuracy is a check on the gain or amplification factor of the vertical deflection system. However, this accuracy is normally checked at a particular fixed frequency. The bandwidth is a measure of the frequency up to which the vertical measuring accuracy can be utilized.

In order to carry out any tests on the oscilloscope, the first requirements are some items of test equipment which are known to be suitable. They must have an accuracy at least 10 times better than the oscilloscope and preferably more. The signals provided by them must be extremely stable and hence maintain their rated accuracies constantly under all conditions of temperature, power supply and so on.

13.1   Internal calibrator

Most modern oscilloscopes have an internal calibrator fitted with an outlet socket on the front panel. It is usually in the form of a terminal or socket to which the probes can be connected. The single calibrator waveform may allow up to four checks to be made, but not all calibrators are so designed, so care must be taken only to use the calibrators for appropriate applications. A calibrator may have the following attributes in order of popularity:

1   *Square wave or pulse type waveform, with medium speed risetime for visible rising and falling edges, and square corners.* The square corners and visible edges allow divider probes to be frequency compensated by adjusting the

probe trimmer for zero overshoot or undershoot (see Chapter 3 on probe compensation). Nearly all calibrators have this facility, and indeed some are called 'probe test' outlets.

2 *Amplitude calibration of waveform.* The above-mentioned square or pulse type waveform is an exact, constant, peak-to-peak amplitude. This output voltage is an exact ratio of the attenuator ranges. For example, the voltage may be 200 mV p.p. So if the ×10 probe was connected on the 5 mV attenuator range, the input sensitivity at the probe tip would be 5 mV × 10 = 50 mV/div. When connected to the 200 mV calibrator outlet, this would produce a vertical display of exactly 4.00 divisions. This calibrator amplitude has an accuracy of better than 1 per cent. Almost all calibrator outlets, as well as allowing probe frequency compensation, also have this exact amplitude calibration accuracy.

3 *Frequency calibration of waveform.* Sometimes, but not always, the output waveform will be a frequency calibrated square wave. The frequency will be stable and independent of temperature and power supply changes, and thus can be used to check the timebase calibration of the oscilloscope. For instance, the calibrator output may be 1 kHz ± 1 per cent. Then on the 1 ms/div range, exactly one cycle per division will be displayed. Not all calibrators have fixed or stable frequency, so check the oscilloscope specification to verify before using the internal calibrator for checking timebase accuracy.

4 *Current calibration of waveform.* A few oscilloscopes feature a calibrator outlet which is at a precise current level, as well as some or all of the above three features. By internally loading the voltage calibrator within the oscilloscope, with a precise load resistance, the calibrator current can be exactly defined. In this case the calibrator takes the form of an isolated metal bar on the front panel of the oscilloscope through which the calibrator current passes. The calibrator voltage is also available from this same point. So if the output voltage was, as above, 200 mV p.p. and the internal load resistance was 40 ohms, then the current flowing through the current calibrator outlet loop is 5 mA p.p.

$$I = \frac{V}{R} = \frac{0.2}{40} = 5 \text{ mA p.p.}$$

A current probe can now be connected round this current conductor to check calibration against a precise 5 mA input.

The specification for a calibrator with all the above features might then be:

> Voltage output   200 mV p.p. ± 1 per cent square wave
> Frequency   1 kHz ± 1 per cent
> Current output   5 mA p.p. ± 1 per cent

Higher performance oscilloscopes often provide extra facilities such as dual voltage and dual frequency outputs. Dual voltage outputs allow a wider range of probes to be checked and compensated, such as ×1, ×10, etc.

Dual frequency calibrators often have an extra high frequency output as well as the standard 1 kHz, or possibly a switch to change from low to high frequency. The high frequency square wave is normally about 1 MHz and allows the optimum adjustment of high frequency probes. These high frequency probes may have two or three compensation adjustments, instead of just the one on standard probes. These probes are set up in much the same way as before, adjusting the corner of the square wave (at 1 MHz) for minimum overshoot, undershoot or ringing. Once the probe is correctly adjusted, it will then have a flat frequency response up to its rated top frequency limit, which may be several hundred megahertz. Of course, high frequency probes can only be adjusted (and used) in conjunction with high bandwidth oscilloscopes.

13.2　Horizontal accuracy

To check the timebase accurately you need a signal with distinct sharp rising and/or falling edges. The objective is to display a waveform of exact known frequency, and align each waveform edge with a vertical graticule line. A rounded waveform, such as a sinewave, would be no good since there is too much error possible aligning each peak with a graticule line. A square or triangular wave is quite good, but best of all is a time mark generator.

Time mark generators are usually controlled by crystal oscillators which give an exact frequency output which remains constant. The output waveform is usually in the form of a sharp narrow needle pulse which is very easy to align precisely with the oscilloscope graticule lines. These generators often provide a range of outputs in a 1, 2, 5 sequence to match the timebase ranges on the oscilloscope. By adjusting the time/division switch on the oscilloscope in conjunction with the time mark generator, it is possible to display one pulse per division on every range of the oscilloscope.

To check a specific range of the timebase, first make sure that the timebase variable control is set to the CALibrated position. Set the time/division switch to 1 ms/div. Apply a 1 kHz marker pulse to the Y input and adjust the amplitude switch and trigger controls to obtain a suitably sized, stable display. Now align the sharp edge of the first pulse on the left, with the first vertical graticule line, and then check the position of the tenth pulse with respect to the tenth graticule line. If it corresponds exactly, then there is no timebase error at all. If the tenth pulse is to the left of the tenth graticule line, then the timebase is too slow. Carefully measure the distance between the tenth marker and the tenth graticule line. (Make sure it is the tenth marker you check. If the error is large it may be more than 10 mm away from the tenth graticule line, and the nearest one may in fact be the eleventh.) Then the error is

$$\frac{\text{Deviation (mm)}}{10 \text{ divisions (100 mm)}} \times 100 \text{ per cent}$$

Let us suppose that the tenth pulse is measured to the left of the tenth graticule line by 2 mm. Then the error on that range (1 ms/div) is

$$\frac{2}{100} \times 100 \text{ per cent} = -2 \text{ per cent}$$

If the tenth pulse had been to the right of the tenth graticule line by (say) 3 mm, then the error would be

$$\frac{3}{100} \times 100 \text{ per cent} = +3 \text{ per cent}$$

To take one more example, suppose the tenth pulse was on the ninth graticule line, the error would be

$$\frac{10}{100} \times 100 \text{ per cent} = -10 \text{ per cent}$$

Now this technique is applied to all the timebase ranges in turn, and every range must be within the specified limits – normally 3 per cent. It is to be expected that most ranges will have some small error, but in every case the error should lie between +3 and −3 per cent (in the case where the timebase accuracy is specified as 3 per cent).

13.3   Horizontal magnifier

If the oscilloscope is fitted with a horizontal magnifier, then the accuracy should be checked with this control in operation. First set the timebase switch to a convenient position, say 1 ms/div, and apply the 1 kHz marker or square wave signal to the vertical input socket. The display should be one marker per division before the magnifier is operated. Now switch in the × magnifier. If the control is say ×5, then each marker will thus be displayed every 5 divisions, and hence three markers should be visible. If it is a ×10 magnifier, then each marker will occur after 10 divisions, so only two markers will be visible, on the first and last graticule lines.

Measure the calibration error as before as the percentage error between where the last pulse occurs and the last graticule line. Note that with a ×5 magnifier, the error between two markers occurs over 5 divisions, however this is the same as the error between the first and third markers over ten divisions.

13.4   Vertical accuracy

The same principles as above are applicable to the vertical measuring system. Once again make sure that the variable controls, in this case the Y VARiable gain control(s) are set to the CALibrated positions. Whether the oscilloscope has one, two or more vertical channels, the following procedures must be applied to each channel in turn. The fundamental sensitivity of the instrument is calibrated on the 'straight through' or most clockwise position of the attenuator, with any gain magnifier controls disengaged ($\times 1$ position).

As with the horizontal system check, a stable signal generator is required – this time of fixed and exact voltage. A square wave generator is commonly used, with the flat tops and bottoms of the square wave ideal for alignment with the horizontal graticule lines. A frequency of about 1 kHz is usually chosen. This frequency is unimportant except for ease of viewing, although the square wave frequency must be much lower than the bandwidth of the vertical amplifier to ensure a faithfully reproduced square wave. Otherwise the frequency should be high enough to view the display without flicker, which occurs at and below timebase speeds of 5 ms/div.

Each position of the attenuator is checked by applying the calibration signal to display an exact amplitude square wave of say 4 divisions. So if the first position of the attenuator is 5 mV/div, then a 20 mV square wave is connected to display exactly 4.0 divisions on the screen. The top and bottom of the square wave are set to coincide with two horizontal graticule lines and the vertical amplifier gain is adjusted for exactly 4.0 divisions. Now if there is an error of (say) 2 mm, that is, the display amplitude is 4.2 divisions, with an input of 20 mV on the 5 mV/div range, then the calibration error is

$$\frac{2}{40} \times 100 = 5 \text{ per cent}$$

Since the displayed waveform is larger than the required 4.0 divisions, the error is +5 per cent. Now for the instrument to be within the tolerance of 3 per cent, the error must be less than 3 per cent of 40 mm which is 1.2 mm.

So checking basic sensitivity, with an input of 20 mV, the observed waveform must lie between 3.88 and 4.12 divisions. Taking the next attenuator position of 10 mV/div, now set the calibration generator to 40 mV square wave and again check that the displayed waveform is between 3.88 and 4.12 divisions. This procedure is then repeated for each position of the attenuator switch, and the whole test repeated with each vertical input of the oscilloscope.

13.5   Vertical magnifier

If the oscilloscope is fitted with a vertical magnifier or gain switch, then you should check the measuring accuracy with the switch in operation.

First set the attenuator to the most clockwise position, say 5 mV/div, then operate the magnifier switch. If it is a ×5 magnifier, then the sensitivity of the instrument will be increased to 1 mV/div. Now set the calibration generator to a suitable output level to give a display of 4 to 8 divisions as before. A square wave input of 5 mV will thus give a display of 5.0 divisions. Check the accuracy as before, a 3 per cent tolerance allowing a deviation of 3 per cent of 50 mm which is 1.5 mm on the 5 division display.

Once the vertical accuracy has been checked, it is necessary to test the frequency range over which this accuracy is maintained.

13.6   Attenuator frequency compensation

As discussed in Chapter 2 on vertical amplifiers, each range of the input attenuator has frequency compensation capacitors associated with it. In order that the oscilloscope has a flat frequency response characteristic from d.c. to its upper frequency limit, it is essential that the frequency compensation trimmers on the attenuators are exactly set. To check this you will need a square wave generator with variable amplitude, and a compensated attenuator probe. Some of the compensation trimmers can be adjusted by connecting a square wave signal directly to the input socket, but there are other trimmers which are only effective when the oscilloscope is fed via a high impedance source. The latter may be a standard ×10 oscilloscope probe, although to check the highest attenuator ranges, say 20 V/div, you will need at least a 200 V square wave. A better system is to use a 2 : 1 compensated attenuator unit, which can be obtained from oscilloscope manufacturers, or made up. This unit consists of a 1 Mohm resistor with an associated parallel trimmer capacitor in a shielded box.

The procedure is to connect the input square wave via the probe mentioned above on the most sensitive range of the attenuator (say 5 mV/div). A square wave signal of about 1 kHz is then applied through the probe (or 2 : 1 attenuator unit) to the vertical input socket. The trigger and timebase controls should then be adjusted to display a few cycles of the square wave signal. The square wave generator should then be adjusted so that an amplitude of about 4 divisions is displayed on the screen. The first essential is to adjust carefully the frequency compensation trimmer on the probe or attenuator unit to obtain a square wave response with no overshoot or undershoot.

Now you can proceed to check the attenuator compensation. Turn the attenuator switch anticlockwise, one step at a time, and increase the square wave generator output level accordingly to maintain a reasonable size display on each range. Every position should display the same square wave response with absolutely no overshoot or undershoot.

If the oscilloscope is a dual or multitrace instrument, each attenuator should be checked in the same way. Before starting each attenuator check, always adjust the probe compensation trimmer on the most sensitive range of that attenuator first. *Do not readjust this trimmer during the checks*, after it has been set on the most sensitive 'straight through' position.

Overshoot or undershoot on any range indicates that the instrument needs calibrating, and does not have a level frequency/amplitude characteristic on that range.

Once the attenuator frequency response has been checked, you can now proceed to check the pulse response and bandwidth of the overall vertical amplifier system.

13.7   Pulse response

In order to check the pulse response of the oscilloscope you will need a pulse or square wave generator with a high quality output signal. The signal must have a frequency of about 1 MHz, and a very fast risetime. The risetime of the square wave or pulse must be at least three or four times that of the oscilloscope vertical amplifier (see Table 12.1). For a 100 MHz oscilloscope, the risetime (Table 12.1) is 3.5 ns. So the generator must have a risetime less than 1 ns. For a 20 MHz oscilloscope, with a risetime of 17.5 ns, the pulse risetime must be less than 5 ns, and so on. The pulse from the generator should have no overshoot or undershoot.

Connect the generator to the oscilloscope and adjust the trigger and time-base controls to display a stable waveform of one or two cycles. Most of the suitable generators have a 50 ohm output impedance, and require a 50 ohm termination at the end of the cable, before going into the oscilloscope (1 Mohm) input. The pulse response of the vertical amplifier will now be displayed as the quality of the rise and square corner of the pulse. There should be no overshoot or undershoot, aberrations or ringing on the waveform. In practice, a small amount of overshoot is usually permitted, but no more than about 3 per cent. This means only a maximum of 1.2 mm overshoot on a 4 division (4 cm) vertical display. Any greater overshoot will indicate a non-linear frequency characteristic, and high frequency amplitude measurements may be unreliable. If the waveform shows undershoot, then the amplifier is not adjusted for its optimum performance, and the bandwidth may be low.

Each channel of the oscilloscope should be checked for a satisfactory pulse response. If the oscilloscope has a good pulse response, the gain/frequency response will be found to be linear and smooth when checking the amplifier bandwidth.

13.8   Vertical bandwidth

The bandwidth of the vertical amplifier is the upper frequency limit. If you connect a sinewave input of (say) 20 mV peak to peak to Channel 1 input, at

1 kHz, it will give a display of exactly 4.0 divisions on the basic 5 mV range. Now if you increase the signal frequency to that of the oscilloscope bandwidth, say 20 MHz, keeping the signal input to 20 mV p.p., the display will no longer be 4.0 divisions, but about 2.8 divisions. So there is a large measuring error at this frequency due to the progressive gain reduction of the amplifier as the frequency increases.

At the oscilloscope bandwidth the measuring accuracy will fall from 3 to 30 per cent:

$$\text{4.0 divisions (at low frequency)} - \text{2.8 divisions (at bandwidth)}$$
$$= 1.2 \text{ divisions}$$

$$\frac{1.2}{4.0} \times 100 \text{ per cent} = 30 \text{ per cent}$$

Now to check the bandwidth of your oscilloscope, you need a variable frequency, *constant amplitude* sinewave generator. First set the generator to a low frequency, say 50 kHz, and adjust the signal amplitude for a display of exactly 4.0 divisions (50 kHz is the standard frequency used in industry). Now keeping the signal generator amplitude setting constant, gradually increase the signal frequency until the display falls to 2.8 divisions ($4.0 \times 0.707 = 2.8$). This will be the -3 dB bandwidth (b/w) point.

It may be more convenient to use other display sizes such as

- 5 divisions (50 kHz) to give 3.5 divisions (at b/w)
- 6 divisions (50 kHz) to give 4.2 divisions (at b/w)

It is not important that the upper frequency limit of the amplifier be *exactly* the bandwidth stated in the specifications, it is the minimum, or bottom limit.

So when checking an amplifier, if you set the generator to the specified frequency limit for the amplifier, for example 20 MHz, the amplitude must be at least 0.707 times the low frequency amplitude. So with a 4.0 division low frequency input, the 20 MHz (say) deflection must be 2.8 divisions or more.

Now if the calibration was exact at the low frequency, on (say) 5 mV range, i.e. 20 mV input gives display of 4.0 divisions (0 per cent error), then the error introduced due to the upper frequency limitation of the amplifier can be determined fairly accurately as follows.

If you are measuring a *sinewave* amplitude at a certain frequency F_m, express that frequency as a percentage of the oscilloscope bandwidth F_b. Then by comparing this ratio with the table below, you can estimate the error at the measured frequency.

Ratio (%)	10	20	30	40	50	80	100 (b/w)	200
Error (%)	0.5	1.5	3	5	10	20	30	75

For example if the oscilloscope bandwidth was 20 MHz, and you were measuring a sinewave at 4 MHz, then the ratio would be

$$\frac{4}{20} \times 100 \text{ per cent} = 20 \text{ per cent}$$

Then from the table, at 20 per cent frequency ratio, the measurement error would be 1.5 per cent.

If the oscilloscope bandwidth was 50 MHz, and the measuring frequency was 20 MHz, the ratio would be

$$\frac{20}{50} \times 100 \text{ per cent} = 40 \text{ per cent (frequency ratio)}$$

Then from the table the measurement error is 5 per cent.

As the frequency approaches the bandwidth of the instrument, the errors get progressively larger so exact measurements are impossible. In general keep the measuring frequency below 20 per cent of the bandwidth to ensure a measuring error less than 2 per cent.

APPLICATION EXAMPLE

Voice recording on a DSO with single sweep

In this example a digital storage oscilloscope (DSO) is used to record a human voice pattern from a microphone. One of the great advantages of a DSO is its ability to capture slow events. Human speech, although containing a wide range of frequencies, consists of collections of slow events, namely words. Even a short word may take a large fraction of a second. By repeating a single word into a microphone and recording the signal into the digital memory, it is possible to look at the voice pattern for that particular word.

Now we have a conflict in the requirements of our digital storage scope. In order to capture a complete word or several words, we need a slow timebase speed so that the duration of the sweep is long enough to capture the whole input from the microphone. However, as the timebase speed is set slower and slower, the sampling rate is also reduced. This is to ensure that the digital memory is always filled in one sweep of the timebase. As the sampling rate is reduced with timebase speed, so the upper frequency limit that we can reliably capture is reduced. As a rule of thumb we can take the upper frequency limit to be about one-tenth of the sampling rate, although in practice this value can be widely varied according to the type of input signal and the application.

So we have a slight conflict between a slow timebase speed to capture the whole event (automatically providing a slow digital sampling rate) and a fast sampling rate to capture the upper frequencies of the input signal. Let us look at some figures.

In Chapter 11 it was explained that the sampling rate, sweep time and memory size are related by the expression

$$N = S_r \times T_s$$

where N = number of samples per sweep.

S_r = sampling rate.

T_s = sweep time.

Let us assume that the DSO we are using has a memory size of $2k \times 8$ bit. So the horizontal memory size is 2k (2000). Therefore, we can assume that in one sweep of the timebase the memory is filled, so the number of samples $N = 2000$. To find the most suitable sweep time for this application we can transpose the formula

$$N = S_r \times T_s \text{ to find } T_s$$
$$T_s = N/S_r$$

We know that the number of samples per sweep is 2000 (N), but we need a figure for the sampling rate S_r.

We will assume that the highest frequency used in human speech is about 4 kHz. In order to capture accurately this frequency we should use a sampling rate about 10 times higher, so let

$$S_r = 4\,\text{kHz} \times 10 = 40\,\text{kHz}$$

Now we can put these figures into the equation $T_s = N/S_r$. Thus

$$T_s = 2000/40 \times 10^3 = 50 \times 10^{-3} = 50\,\text{ms}$$

So if the sweep time T_s is 50 ms, then the sweep rate (for 10 divisions) is 50 ms/10 = 5 ms/div.

So we should use a sweep speed of about 5 m/div for speech capture. In practice, it does not matter if we use slightly lower sweep speeds since we are only interested in the general shape of the waveform, not the exact response of its high frequency content. If faster timebase speeds are used then so much the better to capture all the frequency content more accurately.

First we will set the timebase of the DSO to the range suggested by our figures, 5 ms/div, giving a total sweep time of 50 ms. Initially set the DSO to the required working conditions. Select Channel 1 only,

Channel 1 trigger source, and adjust the trace to the vertical screen centre. Set the trigger controls to normal, a.c. coupled, and select NON-store mode. Set the timebase to 5 ms/div as above. Connect your microphone to the Channel 1 input socket, a.c. coupled, with the screened outer lead of the microphone to ground and the inner 'signal' connection to the Channel 1 input. Set the Channel 1 attenuator switch to the clockwise, most sensitive position. This will most probably be about 5 mV/div.

Now select manual trigger, and if possible adjust the trigger level control to the threshold position where the trace will continuously appear. There may be enough 'noise' present at the input to allow the trace to be triggered and displayed by careful adjustment of the level control. If not, gently tap the microphone while adjusting the level control, so that the trace can be made to appear with the minimum level of input signal. This trigger level point should be about mid-range, and once you have found the correct position, leave the control set to this point.

Now turn the Channel 1 attenuator switch counterclockwise one position at a time until the trace disappears. This is to reduce the level of the input signal 'background noise' going into the Y amplifier to below the trigger threshold level. When you have reached the right position, there will be no trace on the screen. Make a note of this switch position in case you need to return to it later after subsequent alterations of the switch setting.

If there was no trace displayed in the first place due to insufficient background noise to trigger the sweep, then leave the attenuator at its most sensitive position.

With the trigger level, attenuator switch and timebase all set, the next step is to select storage operation. Once in the storage mode you can experiment by speaking into the microphone and watching the results on the screen. Choose one short word and repeat it into the microphone while noting the amplitude and duration of the signal on the screen. If necessary turn the timebase switch counterclockwise one or two steps to ensure that you have captured the whole of the signal within one sweep. If the signal occupies less than about half the screen width, turn the timebase switch clockwise one step at a time while repeating the sound into the microphone. The optimum timebase speed is when the whole signal duration is equal to or just less than the scope screen width.

If your scope has a single shot function in the storage mode, then select single shot and press the reset button. There is usually a lamp on the front panel to indicate 'armed' or 'reset' condition, and this indicator should now be lit.

Repeat your chosen sound clearly into the microphone and watch the results on the oscilloscope. The reset indicator should now be off, and the signal captured on the oscilloscope. If the display is just right for examination, select the hold or save mode to prevent the signal being lost. If the amplitude is too small, release the save button, press reset, and repeat your word slightly louder into the microphone. Once you have achieved a satisfactory display on the screen, select hold or save to secure the signal in memory.

Figures 13.1, 13.2 and 13.3 show the results of three such voice waveforms.

In the first example, Fig. 13.1, the word 'digital' was spoken. Because of the length of this word, it was necessary to reduce the timebase speed to 50 ms/div to capture it all. You can see in the figure that the word digital took about 500 ms or half a second to say. The three syllables: dig-it-al can clearly be seen as three signal peaks. Remember that the trace sweeps left to right, so the three peaks dig-it-al appear left to right on the display.

The second example shows the voice signal produced from the word 'storage'. Since this word is slightly shorter, the timebase speed was increased to 20 ms/div. To record a different sound, release the hold button, press reset, and speak the new word into the microphone. Again it may be necessary to repeat the recording sequence

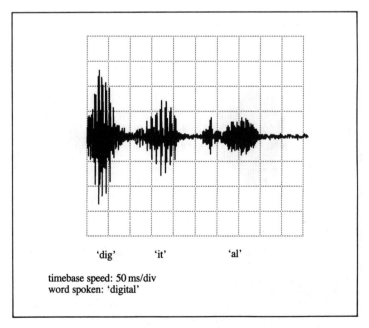

'dig'   'it'   'al'

timebase speed: 50 ms/div
word spoken: 'digital'

Figure 13.1. Screen display of voice pattern of 'digital' captured in digital store.

several times to optimize the timebase speed setting and adjust the volume of sound into the microphone.

If the display amplitude is too small, you should speak louder into the microphone to increase the signal. Avoid turning the attenuator clockwise beyond the optimum switch setting noted earlier as the level of background noise may continually trigger the sweep before the word is spoken.

If the display height is too large, however, there is no harm in turning the attenuator counterclockwise to reduce the display size.

Each time you want to record a new voice pattern, start by setting the attenuator switch to the optimum setting noted above in the original setting-up procedure.

The pattern shown in Fig. 13.2 was produced by releasing the hold button, and alternately pressing reset and speaking the word 'storage' into the microphone. The optimum timebase speed was found to be 20 ms/div, and it can be seen in the figure that the duration of this word was about 10 divisions, which gave a total time of $10 \times 20\,\mathrm{ms} = 200\,\mathrm{ms}$, or about one-fifth of a second.

If you look at the shape of the recorded signal while repeating the word storage into the microphone, you can see the correlation between the variation in the signal and the different sounds that make up the word.

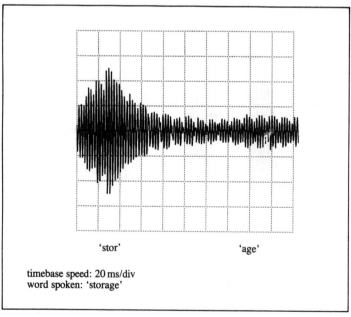

'stor'   'age'

timebase speed: 20 ms/div
word spoken: 'storage'

Figure 13.2. Human voice pattern of 'storage' captured in digital memory.

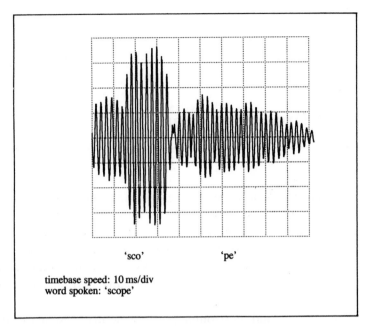

'sco'   'pe'

timebase speed: 10 ms/div
word spoken: 'scope'

Figure 13.3. Stored waveform of human voice pattern of 'scope'.

For a third example the word 'scope' was recorded. Once again, the hold button was released, the reset button depressed to 'arm' the timebase, and the word repeated into the microphone. After each recording of the signal, the timebase and attenuator switches were adjusted to obtain the optimum display before pressing the reset button. Once the best signal was recorded, the hold button was pressed to save the signal, and then the display examined. In this case it was found that the duration of the word was about 100 ms, taking nearly 10 divisions at 10 ms/div.

Comparing the display with the word 'scope', distinct parts of the word pattern could again be seen to correspond with the sounds that form it.

Having followed the above procedures, you can go on to record any sounds you choose. Using faster timebase speeds, single letters or sounds can be recorded and compared with each other and using slower timebase speeds, longer phrases can be tried.

14
First time operation

Before operating an oscilloscope for the first time, make things as easy as possible by setting the instrument into its simplest condition. First, use only one trace if it is a dual or multitrace model and, secondly, ensure that the trigger circuit is in the automatic mode. When the trigger is set to automatic, a trace is displayed on the screen even with no input signal connected. In the manual trigger mode, the screen may remain blank until an input signal is applied.

Adjust the vertical and horizontal position controls to mid-travel and ensure that there are no signals applied to the input sockets. Now switch on the oscilloscope. Adjust the intensity and focus controls to obtain a clear sharp display, and set the timebase switch to 1 ms/div.

Now you need a signal to connect the (Channel 1) input socket. Most oscilloscopes are equipped with a calibration or probe test signal, so this is ideal. Set the (Channel 1) input coupling switch to the AC position and connect a ×10 probe to the input socket.

If there is no signal source available on the oscilloscope, then an external signal generator is required, with a square or sinewave output at about 1 kHz. This signal should then be connected to the Y input using the ×10 probe already connected to the Channel 1 input socket as above.

Some oscilloscopes are fitted with an autoset feature. As the name suggests, this is a system for automatically setting the oscilloscope parameters to optimum to display the applied signal. It will select the trigger source, attenuator range and timebase speed to display a convenient size signal within the screen area. Usually about 2, 5 or 10 cycles of the input waveform will be displayed, at half or full screen height. On some models you can select these parameters in advance, and store them in a memory within the oscilloscope, so that each time you switch on the instrument, you get the type of display you have preselected. Once this is achieved, the operator can take over manual control of the instrument to display the waveform as he or she chooses. If your instrument does not have this autoset feature, then proceed to organize the screen display manually.

Turn the attenuator switch counterclockwise until you can see the whole of the waveform from top to bottom, then set the attenuator to a position to give a vertical display size of about 5 divisions. With the trigger mode set to automatic, the applied signal should already be triggering the timebase, and giving a steady display on the screen. If not, check all the control and switch settings again, especially the trigger functions.

Now assuming all is well, the best way to get to know your instrument is by experience. Taking each section one at a time, check the operation of the three main functions of the oscilloscope: the vertical, trigger and timebase sections. As you investigate the effects of each section, refer to previous chapters to understand the operation of each function. As the behaviour of these functions is interdependent, it is not possible to investigate the three sections completely separately. For instance, if you adjust the vertical amplitude of the signal using the attenuator, it may cease to trigger if the display becomes too small, or even disappear if the trigger is set to the manual mode. So as you are becoming familiar with the oscilloscope, change the settings one at a time, and if something unexpected occurs, go back one step to restore the previous condition.

14.1   Vertical amplifier

Ideally you should have a signal generator available, with variable frequency and amplitude. A sinewave generator is best, although any shape waveform will do. If you do not have a signal generator, there may be an internal calibrator signal available from the oscilloscope itself, but this will often be limited to a fixed voltage and a fixed frequency.

With your signal connected to the Channel 1 input socket, start with all the instrument controls set to their simplest conditions, as above.

Now experiment with the vertical controls to see what happens.

1   Vary the attenuator switch from end to end and use the vertical position control to keep the trace at the centre of the screen (if possible). Try doing this with a.c. and then d.c. input coupling. Set the input signal to give an exact display height, say 5 divisions, and then use the Y variable control and observe its range.

2   If you have an alternative signal source available, repeat procedure 1 with different inputs, such as square waves and pulses. Particularly notice the effect of the a.c. and d.c. positions when looking at square wave and pulse waveforms. When you are happy with the use of the vertical controls, keep a signal connected to the input and set for about 5 divisions display with a.c. coupling on the input switch.

14.2   Trigger controls

The trigger system is the means of displaying or measuring exactly what you want on the screen. The trigger controls allow you to select the part of the waveform from which you want to start the sweep, that is, the point of interest.

First select the manual mode and operate the trigger level control. To see the effects, a slow risetime signal is best – a sinewave or triangular waveform, rather than a square wave.

Try operating the level in conjunction with the polarity switch (+/−) and different trigger couplings. You might find that on some positions, trigger is not possible, such as HF or LF trigger coupling. Refer to your oscilloscope specifications for the limit frequencies and amplitudes for each function. For instance, the trigger circuit may only operate with screen display signals greater than 0.5 division; or on LF trigger, the trigger may only operate on signals slower than say 1 kHz, and so on. Once you are familiar with the specification for each mode, you will be able to use them effectively in future to always trigger successively on any signal.

If you have a source of variable signals, try applying different shape and frequency waveforms to the oscilloscope input, and try triggering on each. You will find that with some fast edge signals, the level control is very sensitive and critical, and with some waveforms the instrument will not trigger on auto at all, but will trigger on manual mode using the level control.

Spend some time to become completely familiar with the trigger circuit as it is the key to successful operation of the oscilloscope. Find out exactly the effect of each control in the trigger system, check its limitations from the instrument specifications and remember its location on the panel layout for future easy access.

14.3 Timebase

The timebase or sweep generator allows you to choose the amount of a cycle, or number of cycles of the input waveform to be displayed. It sets the rate at which the spot travels horizontally across the screen. The scale is in time per division, so the total sweep time for a 10 division horizontal trace, is 10 times the rate. So, for example, on the 10 millisecond per division range of the timebase switch, the total sweep time is 100 milliseconds (100 ms).

Try applying different input frequencies, and adjusting the timebase switch to give displays of one cycle over the whole screen for each input. At low frequencies it may be necessary to use the trigger level control, and possibly LF or DC coupling. Notice the problems of displays below about 10 ms/div. The short persistence of the trace means that it appears as a moving spot, instead of a solid trace.

Note the limitation of the top timebase range which will prevent you from displaying only one cycle of input waveform, once a certain input frequency is reached (assuming the signal generator used will go to a high enough frequency).

Experiment with the operation of the timebase variable control, and the hold-off, then return to the calibrated positions.

Once you are familiar with the operation of the vertical, trigger and time-base functions, you should be ready to start to use the oscilloscope to its full capability. Earlier chapters explain in more detail how each function operates, so a full understanding can be gained of all the systems on your oscilloscope. The only limitations to your use of the instrument will then be the specification limits, rather than lack of operational expertise.

APPLICATION EXAMPLE

Measurement of relay characteristics using a DSO with single sweep

In this example a digital storage oscilloscope is used to capture and measure the characteristics of a low voltage relay. Because the relay is a partly mechanical device, it is inherently slow in operation. However, the examination of slow events is no problem for the DSO. The characteristics that we shall measure are:

1   The switch on time.
2   The switch off time.
3   The relay coil current and voltage.

The relay we will use has an operating voltage of 12 V d.c., and a coil resistance of 100 ohms.

Figure 14.1 shows the circuit we will use to test the relay. Instead of switching the relay directly with a d.c. voltage, a pnp transistor is used. This is to eliminate problems with contact bounce in our operating switch affecting the characteristics of the relay, and also to demonstrate the current and voltage waveforms involved when switching the relay.

The relay is energized by applying a voltage across the coil. The coil has a resistance of 100 ohms in this case, so the d.c. currents involved when the relay is on or off can easily be determined from Ohm's law $(i = V/R)$. However, the relay coil, being wound turns of wire, also has inductance, and this inductance can cause problems for a transistor used to switch the coil. This factor may cause excessive and/or reverse voltages to occur across the transistor, or excessive power dissipation, which may cause damage or even total destruction of the transistor if not catered for in circuit design.

The first measurement we will make is the relay turn on time. With the switch connected to the +12 V supply as shown in Fig. 14.1, the transistor is switched off, and hence the relay is also turned off. The relay contacts are also connected between the +12 V supply and

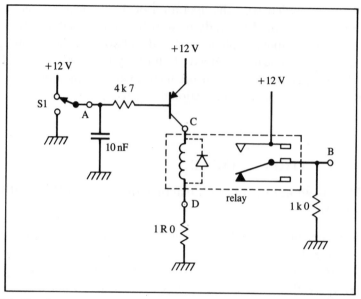

Figure 14.1. Circuit configuration for testing a relay, with probe connection points.

ground as shown. With the relay off the 1 kohm resistor holds the relay contact at ground potential, and when the relay is turned on, the +12 V contact takes the resistor potential up to +12 V.

To measure the 'turn on' time of the relay, we will connect the Channel 1 oscilloscope probe to the switch wiper, point 'A', and use the switching voltage as the trigger and time reference connection. So connect the Channel probe to the point 'A' in Fig. 14.1 and set up the required scope conditions. All the waveform measurements are made relative to ground, so in each case the probe earth clip should be connected to the earth point in the circuit.

Select analog (non-store) dual trace, chop mode and also select Channel 1 trigger source. Since point 'A' is initially at +12 V potential and will be switched to ground potential to turn on the relay, this is a negative going voltage change, so select negative slope on the trigger polarity selector switch on the scope. Assuming a screen height of 8 divisions, we will keep the Channel 1 display to less than 4 divisions to avoid the two traces overlapping. So with a ×10 probe from point 'A' to the Channel 1 input socket, select the 0.5 V attenuator position (this gives a sensitivity of 5 V/div at the probe tip). Now with a 12 V input signal and 5 V/div we will get a display amplitude of $12/5 = 2.4$ divisions. The input coupling switch (a.c.d.c.GND) must be set to d.c.

The Channel 2 ×10 probe is then connected to point 'B', the relay output contact. Note that there is no electrical connection between

the relay coil circuit and the output contacts. Since the second input
will also have an amplitude of 12 V, also set the Channel 2 attenuator
switch to 0.5 V/div (assuming a ×10 probe is used) and the input
coupling switch to d.c.

Now select manual trigger and repeatedly operate the switch S1 to
energize (and de-energize) the relay. Keep adjusting the trigger level
control until the sweep fires on every 'on' cycle of the relay. Now
select store mode, single shot, and press reset. If you have a pretrigger
facility on your instrument, select about 10–25 per cent pretrigger at
this point.

Leave the level control set to the optimum position found above,
and still repeatedly switching the relay, adjust the timebase switch to
best display the relay contact close point. If you are using the single
shot facility, after each cycle, press the reset button to arm the timebase for the next sweep. If necessary adjust the two vertical position
controls so that both levels of display on each channel are conveniently placed on the screen.

The relay contacts closing results in the Channel 2 trace going from
ground to +12 V. Figure 14.2 shows the results of this test. The start
point of the sweep on the left (the trigger point) is where the switch
(S1) contacts close. In this display, since pretrigger was not used, you
can just see the negative excursion of the switch wiper voltage from

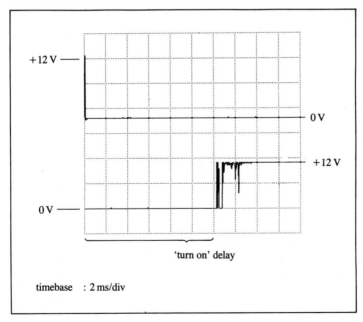

Figure 14.2. Screen display of waveform from relay contacts when relay is energized.

+12 V to 0 V on the left. After a slight delay, the lower trace shows when the relay contacts close. In this example the delay can be seen as 6 divisions at 2 ms/div, giving a 'turn on' delay of 12 ms. The 'noisy' waveform at the transition point from 0 V to 12 V on the lower trace is caused by contact bounce on the relay contacts. After about 2 ms, the voltage settles at the +12 V level.

The second test is the relay 'turn off' time. Using exactly the same conditions as before, it is now a matter of capturing the other phase of the relay action, when the switch, the transistor, and hence the relay are turned off. The starting point will be with the switch wiper grounded, the transistor conducting and hence the relay energized. The switching action monitored by the Channel 1 probe will now be positive going (0 V to 12 V) so we must reset the trigger conditions.

First release the store mode, and revert to analog operation. Select positive trigger on the trigger polarity (+/−) selector, and again repeatedly switch S1 while adjusting the trigger level control. Set the trigger level so that the sweep fires on every 'switch off' cycle of the relay when the switch wiper (S1) goes positive. Next select storage, single shot and reset as before. Keep switching the relay off (and on) and adjust the timebase switch for the best display of the turn off delay time. In Fig. 14.3 you can see the results of this test. The delay

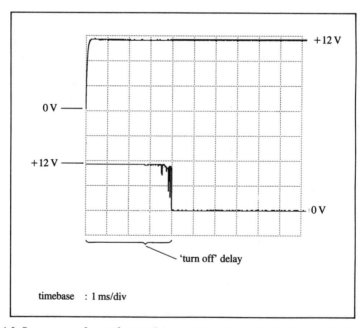

Figure 14.3. Screen waveforms from relay contacts when relay is de-energized.

time from the trigger point at the start of the trace to the relay turn off is only 4 divisions. With the timebase set to 1 ms/div, this gives a delay of 4 ms.

So for this relay, the delay time to turn on is 12 ms and to turn off 4 ms. The difference is largely caused by the spring holding the relay contacts which is normally off when the relay is de-energized. When the relay turns on it must operate against this spring (taking longer) whereas when it turns off, it is assisted by the spring action.

The third test is to measure the voltage and current in the relay coil. You can measure these waveforms for both the on and off cycles of the relay, but it is most instructive for this example to look at the turn off condition. Release the hold, single shot and store modes, and revert to the analog condition. Transfer the Channel 1 probe to the relay coil, point 'C' in Fig. 14.1. When the relay turns off, the transistor collector point 'C', will go negative to ground potential, so again choose negative slope on the trigger polarity selector. Transfer the Channel 2 probe to point 'D'. This point is across a 1 ohm resistor in series with the coil. The resistor conducts the same coil current, so the voltage developed across this resistor is directly proportional to the coil current.

From Ohm's law $I = V/R$, since $R = 1$, here $I = V/1$ so $I = V$. So for every 1 volt across the resistor, there is 1 amp flowing through it. Or more practically, for every 100 mV across the resistor, there are 100 mA flowing through it (and the coil). Starting with the relay in the 'on' position, the transistor collector, point 'C', will be at a positive potential. With the relay coil passing current, which also flows through the 1 ohm resistor, a positive voltage will also be developed across this resistor.

Again using Channel 1 as the trigger source, with negative trigger selected, adjust the trigger level control while repeatedly operating the switch S1 until the timebase triggers on every turn off cycle of the relay. Then select store, single shot and reset. Continue to switch S1 while adjusting the timebase switch for the best display. Remember to operate the reset button after each cycle. It will also be necessary to adjust the Channel 2 attenuator setting during this operation to optimize the display of the current waveform in the coil. Once you are happy with all the switch settings, select the hold or save function to secure the waveforms in the memory.

Figure 14.4 shows the voltage and current waveforms during switch off. The upper trace shows the relay coil voltage changing from about 10 V to 0 V over a period of 8 ms. However, you can see that as soon as the transistor is switched off, the collector goes negative to about −20 V before gradually rising to ground potential. This negative

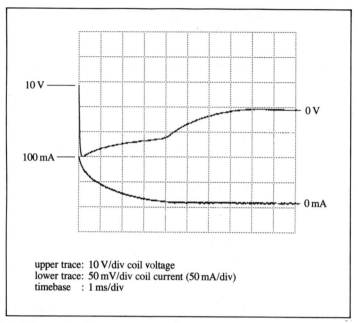

Figure 14.4. Relay coil voltage and current waveforms when the relay is de-energized.

excursion, caused by the coil inductance, can destroy the transistor unless guarded against. At the point of switch off there are more than 30 V across the transistor. To avoid this problem, a diode is normally fitted across the relay coil (shown dotted in Fig. 14.1). If the transistor collector goes negative to ground, the diode conducts and then clamps the collector at ground potential. This limits the voltage across the transistor to about 12 V, instead of the 30 V peak voltage measured in this example. You can see from these waveforms that the relay currents and voltages do not simply turn on and off as the switch is operated.

The coil inductance causes a longer time constant to be introduced into the circuit, and the current and voltage to be out of phase with each other. Due to the many variations in relay construction such as coil resistance, inductance, size, spring strength and so on, the characteristics of relay operation are widely varied. The above technique can be used to select the most suitable relay for a particular application once the main parameters are known from its specification.

15
Interfacing to other equipment

Before the introduction of digital storage oscilloscopes, the output facilities available on oscilloscopes were usually limited to four types of signal: vertical signal output, plotter output, sweep ramp output and gate output. These signals could be fed into other equipment either as control or timing references for that other equipment, or for further examination. All these features are still commonly available on the current range of analog oscilloscopes and have many applications.

15.1 Vertical signal output

This signal is an amplified version of the main signal entering the Channel 1 or Channel 2 input socket. It may be selectable from either channel, or possibly direct from one channel only. It is a buffered output usually with a fixed ratio to the input signal, say 50 mV/div. That is 50 mV p.p. signal output for each division of signal displayed on the screen, irrespective of the attenuator setting. So if the input attenuator is set to the 5 mV/div position, then the vertical output level is 10 times the input signal level. If the input attenuator is set to the 50 mV/div range, then the output signal is the same amplitude as the input signal. The fact that it is a buffered output means that it is driven by a low impedance source, capable of feeding into other equipment without affecting or loading the oscilloscope. This signal can be fed into other equipment such as a frequency counter, distortion meter and so on, or possibly back into the other input channel of the oscilloscope. This is a method of increasing the sensitivity of the oscilloscope by using one channel as a pre-amplifier for the other. In the example above, where the input sensitivity of the oscilloscope is 5 mV/div, and the vertical output signal level is 50 mV/div, the system gain is 10 (50 mV/div divided by 5 mV/div). An input signal could be fed into Channel 1 (set to 5 mV/div); the 10-fold magnified output taken from the vertical output socket and fed back into Channel 2. The Channel 2 display would then have an effective sensitivity of 5 mV/div multiplied by $10 = 500 \mu$V/div. Another use for the vertical output is to feed into a plotter, to obtain a hard copy of the displayed waveform. These plotters may have two methods of operation known as Y–T and X–Y.

The Y–T version uses the vertical (Y) signal from the oscilloscope to drive the pen on the plotter in the vertical axis. The Y–T plotter has an internal timebase (T) or paper drive to drive the pen or paper in the horizontal axis.

The X–Y plotter takes both its vertical and horizontal drive signals from the oscilloscope, the vertical output as above, and the horizontal from the oscilloscope timebase ramp output socket.

15.2   Plotter output

Some oscilloscopes are fitted with a specific plotter output connector. This is usually a connector with a minimum of six pins supplying the three plotter signals and their respective earths. These are:

> Y output – for vertical analog pen drive.
> Y ground – signal ground for above.
> X output – sweep ramp for horizontal pen drive.
> X ground – signal ground for above.
> Pen lift – TTL command to lower pen during write, and lift pen during flyback, etc.
> Ground – earth for above or cable screen.

This plotter output is a combination of the ramp and vertical outputs referred to above, but with the addition of the pen lift control and the benefit of one multipin connector to feed direct to the plotter through one cable.

In the case of digital storage oscilloscopes, the plotter output signals may be retrieved from the digital memory. In this case digital-to-analog converters are used to transfer the digital stored signal in memory to an analog drive signal for the plotter vertical input. An associated ramp is also generated for the horizontal drive, synchronous with the vertical signal conversion, and a digital pen down/up command is output at the beginning and end of the signal output sequence.

15.3   Sweep ramp output

This output is a buffered version of the ramp waveform from the oscilloscope sweep generator. It can be used as above for the horizontal drive on a plotter, or for other applications requiring a voltage ramp waveform synchronous with the oscilloscope's horizontal scan. An example of this is the f.m. (frequency modulation) input of a wobulator or oscillator. The (ramp) voltage change on the input to the oscillator causes its frequency to change, so that the oscillator output is a swept band of frequencies, synchronous with the oscilloscope sweep. If this swept frequency signal is fed into the oscilloscope (vertical) input, the display will be of amplitude versus frequency (not time) with the lowest frequency on the left, and the highest frequency on the right.

If now a passive filter is placed in the signal path between the oscillator output, and the oscilloscope vertical input, its characteristic frequency response is displayed on the screen.

The amplitude of this sweep ramp output signal is about 5 V, usually starting at zero or 0 V when the sweep starts on the left, and rising to +5 V at the end of the sweep on the right.

15.4   Gate output

The gate output signal, as the name suggests, is provided for 'gating' other equipment. 'Gating' means opening or closing a gate, or switching on or off some circuit or equipment. The gate output signal is synchronous with the oscilloscope timebase ramp, but unlike the ramp waveform, it has a pulse format. That is, there is a sharp transition (fast rise) when the sweep starts, a flat or level output during the sweep time, and a sharp transition (fall time) when the sweep ends. The gate output is usually at TTL (0–5 V) level. The fast edges at the beginning and end of the pulse can be used to switch, or 'gate' other equipment, exactly at the time the oscilloscope sweep starts and ends.

The development of the digital storage oscilloscope meant that the analog vertical input signal was transformed by means of an analog-to-digital converter, into digital format for loading into a digital memory. Having this digital representation of the input signal, it was soon realized that this format enabled easy data transfer from the oscilloscope to other equipment. The way that the data are transferred takes two forms:

1   The first is a dedicated 'bus' system, which is peculiar to each manufacturer. These dedicated systems employ a variety of plug, socket and cable types to transfer data, mainly into hard copy printers, built into; fixed on to; or separate from the oscilloscope itself.
2   The second form of transfer is via a standard 'bus' system, with fixed specifications, and the two in common use are the RS232C serial bus, and the IEEE488 parallel bus. These bus systems allow data transfer between the oscilloscope and a whole range of other equipment including plotters, printers, computers, multimeters, analysers, and so on. They use (almost) standardized plugs, sockets and cables, and enable many different manufacturers' instruments to be interfaced to each other.

15.5   Dedicated systems

There are many oscilloscopes on the market with associated equipment available as accessories or options. The hard copy printer is becoming a standard option, and oscilloscope manufacturers usually provide all the interconnections necessary to link it to the oscilloscope. The object is to obtain a permanent record of the waveform on the screen, and many printers include the measurement parameters (T/div, V/div, etc.) as well as the waveforms.

There are several 'oscilloscopes' available now with a hard copy printer built in. With increasing levels of safety and quality standards in manufacturing, test and calibration, the hard copy printer provides verification of tests carried out on equipment. Anyone wishing to test, purchase or approve equipment has a visual record of its performance.

As well as hard copy printers, a few manufacturers provide further systems which integrate with their oscilloscopes. These often include computer systems with a monitor, disk drives, and some form of data transfer system. Being dedicated systems, they are supplied with all interconnection hardware, operation information and sometimes software support. However, these dedicated systems are in the minority, most oscilloscope manufacturers providing a standard output socket only. Users can then utilize standard bus systems to control their instruments using mainframe or personal computers (PCs). Although almost any type of computer can be used, with appropriate hardware, we shall look here at the most common controller, the PC.

Data transfer systems using standard interfaces such as the RS232C and IEEE488 utilize three types of instrument on the data link. The RS232C was originally designed for two units only, one sender and one receiver, while the IEEE488 was designed for up to fifteen units linked together.

These are controllers, talkers and listeners. Obviously the minimum requirement is two: a talker and a listener. In this case, one must also be a controller. With multi-instrument systems, there may be many talkers and listeners all joined together, and their function at any one time is determined by a controller, which may be one of them. At any given time there may be many listeners active, but only one talker and one controller.

With sophisticated control software it is possible, for example, for data to be output from the oscilloscope (or other equipment) at timed intervals, or when a new signal enters the oscilloscope, or both. Thus automated signal acquisition, measurement and storage is possible using these bus systems.

15.5.1 RS232C

The RS232C is a serial interface standard. Serial transfer means that the data are transferred one bit at a time. For example, if a waveform was stored in the oscilloscope's $1k \times 8$ memory, the signal is encoded as an 8 bit (vertical) by 1k (horizontal) word pattern. There are 8 samples of vertical information for each one of the 1000 horizontal increments across the screen. The serial interface transfers these 8000 bits of information from the oscilloscope to the computer one after the other, until all 8000 have been sent.

The RS232C standard specifies a maximum of 25 connections, but in practice, oscilloscope interfaces only require a few of these, so a 9 pin 'D' connector is often fitted. Because of the two 'standard' plug and socket systems, a 25 to 9 pin adaptor is readily available from computer product suppliers.

The 'D' connector is a multi-pin plug and socket system with a 'D' shaped outer shell. This 'D' shape is for polarization purposes, ensuring that the plug can only enter the socket one way round. The interface allows signals to be passed in both directions.

Although most oscilloscope manufacturers only allow data to output from the oscilloscope to other equipment, a few do permit data to be returned back into the oscilloscope, and the waveform displayed on the screen. Control signals pass both ways on all models, since some sort of recognition is required at each end of the data link to 'know' when to start and stop sending data. This is known as handshaking. There are four types of lines used in the interface: data, control, timing and ground. In general, data lines only carry data from the oscilloscope to the other equipment, whereas control signals pass both ways. Some sophisticated instruments allow remote parameter control using these lines, so that the oscilloscope front panel settings can be changed remotely by a computer for instance. This then allows the possibility of automatic signal measurement, data plotting and automatic test systems. However, the difference should be noted between oscilloscopes which can be commanded to send data, and reset to load a new signal into memory; and those which can have their measuring parameters (volts/div, time/div, etc.) remotely changed by the controller.

The voltage levels specified for the RS232C lines are *not* TTL and may be up to plus or minus 25 V.

The speed at which data can be transferred is known as the baud rate. This is the number of events per second that can be transferred. It may (or may not) be the same as the bit rate, and for oscilloscopes is usually set between 1200 and 9600. It is normally possible to set the baud rate on a switch on the oscilloscope or interface, and it may be necessary to set the transfer mode by means of another switch. Due to the simplicity of the RS232C system, and its widespread use on computers, printers, etc., it is often possible to direct link an oscilloscope to a printer, for instance a 9 pin or 24 pin standard computer printer. Then just one button is required to transfer the data to the printer and obtain a hard copy.

15.5.2 IEEE488

This is a parallel data interface and is almost the same as the IEC625 interface. It was developed from the Hewlett-Packard Interface Bus (HPIB) and standardized as an instrument interface, also known as the General Purpose Interface Bus (GPIB). It uses a 24 pin 'D' type connector, and allows up to 15 instruments to be simultaneously connected together. In order that several instruments can be connected, each instrument on the bus may require at least two connections to it: one from the previous instrument in the chain and one to the next. To allow this to happen, two kinds of IEEE connectors are available: a straightforward 24 pin plug with securing screws; and a 'piggy back' version with an integral socket opposite the plug. This 'piggy back' version is a plug and socket together, the socket being mounted behind the plug, so that when it is connected to an instrument, another plug can be connected into it to carry the bus cable on to the next instrument in the chain. Standard IEEE cables are a maximum of 2 m long, this being the maximum specified for this bus system. Connection of longer cables may prevent the system from working properly due to the extra capacitances of the cables, which the drivers may not be capable of supplying.

The system specifies eight control lines, eight ground lines and eight data lines. The parallel system allows 8 bits at a time to be transferred through the eight data lines, and thus is much faster than the serial RS232 system. Data transfer speeds up to 1 Mbyte/s are possible. Since it is a parallel system, each clock cycle transfers 1 byte (8 bits) of data. So the 1 Mbyte transfer rate means that 8 million bits of data can be transferred per second.

The three types of device that can be connected to the bus in the IEEE488 system are talkers, listeners and controllers.

- *Talkers* can transmit *data* only.
- *Listeners* can receive *data* only.
- *Controllers* can transmit *commands* and transmit/receive *data*.

There may be many listeners active together, but only one talker and one controller at any one time.

Oscilloscopes are normally talkers, although some sophisticated models may also be listeners too. Printers are normally listeners only, and computers may be controllers, listeners and talkers.

To ensure that only one talker is active, each device is assigned an address, and this address is activated by the controller. Complex arrangements for gathering information from different talkers can be made, with the controller (usually a computer) making each talker active when required.

Status information gathering from each talker is known as polling and the talkers can be checked by serial or parallel polling. The status can then be determined to find if the talker is ready to send data or not, and its status sent back to the controller as a 1 bit or 8 bit code. With parallel polling, a maximum of 8 devices can be polled, each able to give 1 bit status reply. With serial polling, all devices can be polled and each can give an 8 bit status reply, but consequently is slower.

Different arrangements can be programmed such as timed data sampling, data transfer after signal threshold changes and so on, and the oscilloscope can be reset by the bus to refresh its memory with new data (input signals).

All data and control signals on the IEEE bus are TTL level, and allow for high speed data transfer by different methods such as one byte at a time, or block transfer, where say 1k bytes are transferred together.

Most personal computers (PCs) are fitted with the RS232 serial port, but it is necessary to add an interface card to the PC to allow it to operate on the IEEE bus. These IEEE cards are fitted into one of the spare slots in the PC, and have the 24 way connector mounted on the card, and accessible at the rear of the PC.

It is essential to set the switches on the card prior to installation, to give the card an address so that the computer can 'recognize' the card and operate it. Consult the card suppliers' instructions for the correct setting of these switches.

IEEE bus operating programs (software) are available from the oscillo-scope manufacturers or the IEEE card suppliers. These will allow the user to operate instruments on the bus without any knowledge of the IEEE bus commands, or having to write their own programs. In fact 'user friendly' software is available to load into the PC and prompt the user via screen menus. Then the operator has only to choose from the options presented on the screen and the computer will do the rest.

Once the oscilloscope waveforms have been transferred to a computer, a variety of operations can be carried out, depending on the software options. Typical choices are disk storage of data; VDU waveform display; mathema-tical waveform analysis; measurement; magnification, printing, etc. Also, as mentioned above, the data can form part of an automatic test system for other equipment.

Figure 15.1 shows the typical configuration for an oscilloscope/controller interface. The connection bus can be either RS232C or IEEE488, and the pin connections are shown in Fig. 15.2. Standard interface cables are normally used for interconnection, so pin identification is taken care of automatically.

APPLICATION EXAMPLE

Use of a DSO to isolate mixed frequency signals

A digital storage oscilloscope can often be used to great advantage when triggering difficulties prevent a stable waveform display from being obtained on the screen. In this example a DSO is used to examine a signal which is a combination of two frequencies added together.

Figure 15.3 shows the signal we wish to investigate. At the timebase speed shown in the figure, 5 ms/div, both the sinewave signals in the waveform can be clearly seen, one much faster than the other.

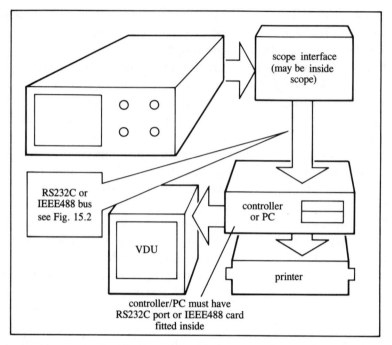

Figure 15.1. Block diagram of interface bus system.

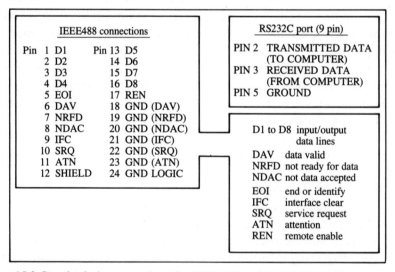

Figure 15.2. Standard pin connections for IEEE488 and RS232C interface connectors.

Suppose that we want to measure the frequencies of both these sine-waves, and therefore wish to obtain a stable triggered display of each on the screen. It may be possible to use the trigger filter to discriminate one frequency from the other as they are fed to the trigger

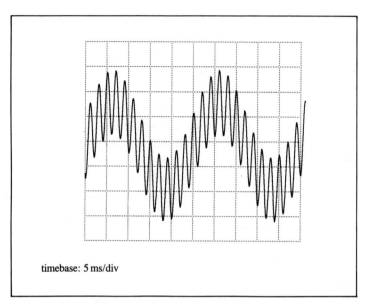

timebase: 5 ms/div

Figure 15.3. Screen display of mixed frequency waveform.

circuit. If that is the case, it may not be necessary to use the storage mode at all. By using LF (low frequency) trigger coupling on the trigger selector, the lower frequency signal in the waveform may possibly be locked, while the higher frequency signal 'runs through'. Then, using the HF (high frequency) trigger position, the other, higher frequency content may be triggered enabling the period to be measured.

With the waveform shown in Fig. 15.3, this trigger filter technique was tried, to alternately lock each frequency so that its period could be measured. However, it was not successful. In either position of the trigger selector, both the frequencies in the waveform seemed to 'run through' on the screen. This was because the two frequencies were too close together for the filter to discriminate. For the trigger filter system to be successful, one of the input signals must fall within the frequency range of the filter, while the other lies well outside it. For instance, the LF position of the trigger filter may pass signals up to 10 kHz, and reject frequencies above. If one of the frequencies was (say) 5 kHz, and the other 50 kHz, then the filter would allow the lower 5 kHz signal to trigger but not the higher frequency. But in this case, where both frequencies in our waveform lie within this range, we cannot use the trigger filter to separate them. This is where we can use the storage system to good effect.

Turn the timebase switch to a slower position where you can see that several cycles of the lower frequency signal are displayed on the screen. Although there is the higher frequency signal superimposed, ignore these faster peaks and look for the concentration of lower frequency peaks on the screen. Use the trigger controls to obtain the most stable display possible. The required coupling will depend upon the input frequencies you are using. In this example, the LF position was found to give the best results, together with manual trigger and careful adjustment of the trigger level control.

Now select storage mode and single shot, and then press the reset button. Look at the results displayed on the screen. If possible try to align the first lower frequency waveform peak on the left with a vertical graticule line using the horizontal position control. If the display is not suitable to do this, press reset again and check the new display. Each time you press reset, the display may change slightly due to the change in trigger point caused by the problems encountered earlier. Now we can use this change of trigger point as a benefit to allow us to move the display along each time we press reset until the first peak aligns with the first graticule line. Once you have achieved the best display possible, select hold or save to secure the waveform in the memory. Again use the horizontal position control to finely align the first low frequency signal peak with the first graticule line. In Fig. 15.4 you can see that there are eight cycles of the lower frequency signal displayed at a timebase speed of 20 ms/div. So the total sweep time is 10×20 milliseconds $= 200$ ms. Therefore, the

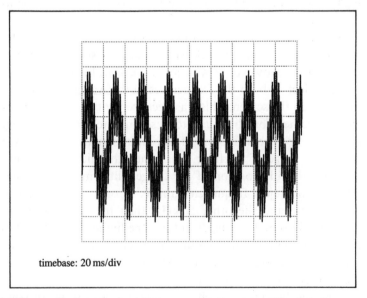

timebase: 20 ms/div

Figure 15.4. Screen display of mixed frequency signal, captured in digital memory for lower frequency identification.

period of the lower frequency signal is 200 milliseconds divided by eight cycles, that is 200/8 = 25 ms. You can measure off the period of one cycle of the lower frequency signal directly from the graticule, but because of the many cycles of high frequency superimposed, it is difficult to see exactly where the beginning and end of each cycle occur. Thus it is more accurate to measure over the largest amount of full cycles possible on the screen. The frequency is the reciprocal of the period so

$$f = 1/25 \times 10^{-3} = 40\,\text{Hz}$$

Next we shall use a similar procedure to measure the higher frequency signal period. Release the save or hold functions, single shot and storage, and revert to analog mode. Turn the timebase switch to a faster setting so that several cycles of the high frequency signal are displayed. Use the trigger filter and level control as before for best display of the signal, and if necessary readjust the timebase speed. Do not worry if you cannot get a good triggered display as again we shall use the storage mode to overcome this problem.

Once you have the best result possible, select storage, single and reset as before. Use the same technique of reset and check until you get a good display of the high frequency signal with the first peak aligning with the first graticule line. Change the timebase setting if required and use the horizontal position control so that you achieve a

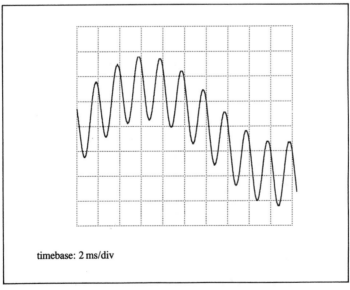

timebase: 2 ms/div

Figure 15.5. Screen display of mixed frequency signal, captured in digital memory for higher frequency identification.

suitable aligned display of several cycles of the signal. Then select hold or save to secure the signal and make your measurements. In this example, Fig. 15.5 shows that the higher frequency signal occupied exactly 10 divisions for 10 cycles of the waveform. The timebase speed was 2 ms/div, so the period of the signal was 2 ms. Taking the reciprocal as before, the frequency is thus $1/2 \times 10^{-3}$. Therefore $f = 500$ Hz.

This technique of 'freezing' a signal on the screen can be used in many applications where triggering is difficult. By simply storing a signal that is rolling across the screen, careful examination is possible, just as when the signal is triggered solidly.

16
No display – what to do

One common problem with oscilloscopes is the total loss of picture on the screen. In most cases this is simply due to the front panel controls being set to the wrong conditions. When the trace has disappeared, we can divide the causes into four main categories:

1 Brightness or intensity (Z axis).
2 Y or vertical deflection.
3 X or horizontal deflection and trigger.
4 Fault conditions.

It is not too difficult, especially with a complicated oscilloscope, to 'lose' the trace altogether. So when this happens it is advisable to check methodically through the control of each of the X, Y and Z axes before assuming the fourth option – a fault condition. The checks should be carried out in the order above: 1, 2 and 3 to ensure success in finding the trace as soon as possible. If the trace suddenly appears due to adjustment of any control or signal during the checks, then consider fully the effect of that control or signal since it is causing the loss of trace. If the problem is identified, appropriate steps can be taken to retain the trace on the screen, and then further checking is not required.

Some oscilloscopes are equipped with indicators to help to find a 'lost' trace. These are called various names according to what they do, such as beam finder, trace locate and overscan indicator. They are self-explanatory, and in some way indicate the direction of the lost trace. These facilities should be used before or in conjunction with the following checks to find the cause of the problem.

Since the most common cause of loss of display is the trigger setting, it is worth checking this first. Look at the trigger control area of the front panel and set the AUTOMATIC/MANUAL trigger selector to the automatic mode. In many cases this will return the display immediately as the sweep can now run whether or not an input signal is present. Even if the display still remains elusive, leave the trigger set to the automatic mode and proceed with the remainder of the checks.

16.1   Brightness

Turn the intensity control to maximum (temporarily). If the intensity control is set too low to see the trace, we have no hope of finding it. If there are any other factors affecting trace brightness, these should be tried as well. If a Z modulation input is being applied, then try removing it. If there is a differential brightness control, set it to maximum intensity.

Has the trace now appeared? If it has, due to increasing the brightness, then there is no further problem. If it has appeared due to an external Z modulation signal being removed, then look into the amplitude, d.c. content, frequency and polarity of this signal, and check against the oscilloscope specifications to see why it is cutting off the beam. Then take steps to prevent the effect recurring. If there is still no display visible, leave the intensity set to maximum and proceed to the next step.

16.2   Vertical (Y) amplifier

If the oscilloscope is a dual or multitrace model, set the vertical mode control switches to display one channel only. If signals are being applied to both channels, display only the channel from which you wish to trigger.

Now first set the vertical position control to mid-travel and the input coupling switch to a.c. in case the trace is positioned just off the top or bottom of the screen. The trace may have appeared if the coupling switch was moved from d.c. to a.c. coupling. This indicates that the input signal contains a high d.c. content which shifts the trace off the top (positive d.c.) or bottom (negative d.c.) of the screen. Next turn the attenuator switch counterclockwise one step at a time. If the waveform was too large for the screen display size, then upon turning the switch, it should come into view. If there is still no trace on reaching the last position of the switch, leave it set there and change the input coupling switch to the ground (GND) position. Then when the trace is eventually displayed it should be a straight line across the (vertical) centre of the screen. If there is still no display, proceed to the next step.

16.3   Timebase and trigger

First make sure that the oscilloscope is set to timebase sweep mode, so check that it is *not* set to XY, horizontal external, component tester, single shot or any other non-basic sweep mode. If there is a sweep delay system or dual timebase system, switch to main timebase or A only mode, so that the delay facilities are disabled. If the trace returns due to the switching of any of these alternative operating modes, the problem may simply be that you are in an undesired mode, or if in the delay timebase mode, that other conditions are wrong. For instance the delayed timebase speed may be too fast or the delay time too long. If there is still no sign of the trace, the last thing to try is the triggering. First make sure that the trigger is set to AUTO mode and not manual trigger. This should ensure the continual running of the sweep in the absence of an incoming trigger signal.

Set the trigger level control to mid-position or to the auto position if on the same control. Switch the trigger coupling to Channel 1 or Channel A trigger source, a.c. coupling.

By now you should have a trace on the screen. Each of the above steps should be carried out slowly, so that when your trace reappears, you know the reason for it. If at the end of this check there is still no trace visible, it is worth repeating the previous two checks.

16.4   Fault conditions

If the three checks above have been carried out methodically and there is still no trace visible, it is almost certain that the instrument is faulty, and in need of repair by appropriate qualified personnel.

Modern oscilloscopes do not 'drift' very much, that is, they do not go out of calibration, or alter the trace positions or indeed the overall specification by any significant amount. Any changes due to ageing, or environmental effects such as temperature, humidity, voltage supply and so on, would only be of the order of 1 or 2 per cent, and not enough to take the instrument outside its calibrated specification. Consequently, if any calibration errors are found, or deterioration in performance, it is almost certainly due to a fault condition, and cannot be remedied by adjustment of any preset control, either internal or external. Under these conditions the instrument should be repaired by qualified personnel, preferably the manufacturer.

APPLICATION EXAMPLE

Measurement of a signal with d.c. offset

A repetitive symmetrical signal may have no d.c. offset relative to zero potential. Each excursion of the signal has an equal positive and negative peak value relative to zero, and the total peak-to-peak value of the signal can easily be measured on either the a.c. or d.c. setting of the input coupling switch. In this example, however, we shall measure the amplitude, d.c. offset and frequency of a signal that is not at zero potential, and use the cursor measurement system for accurate results. The signal we shall look at is a sinewave with d.c. offset.

Before connecting the input signal, first select Channel 1, single trace operation on the oscilloscope, with auto trigger and the vertical input coupling switch set to d.c. Set the Channel 1 attenuator switch fully counterclockwise and the X and Y VARiable controls to their

CALibrated positions. Adjust the Channel 1 position control to set the trace at the vertical screen centre and then connect the signal to the Channel 1 input socket. Turn the attenuator control clockwise until the display is just contained within the screen area at the top or bottom of the screen. Now the important thing is to note which way the trace has moved. If it is still symmetrically disposed about the screen centre, then there is no d.c. offset, but in this example the display at the top of the screen indicates a positive d.c. offset. Note the direction of the offset. If upward (as here) the baseline will need to be set at the bottom of the graticule, whereas if downward, the baseline must be set at the top.

So now with the positive offset in this example, *set the input coupling switch to ground,* and use the Channel 1 position control to align exactly the trace with the bottom graticule line. (If the d.c. offset is small compared to the peak-to-peak signal amplitude, it will be better to set the trace to a line above the lowest graticule line.) If the d.c. offset is negative, then at this point the trace should be set to the top graticule line, but again, if the d.c. offset is relatively small, set the trace to a line between the top line and the centre line.

Take care not to move the position control again after it has been set to the baseline and once again set the input coupling switch to d.c.

Keeping the Channel 1 Y VARiable control set to the CALibrated position, adjust the attenuator switch to the most clockwise position where the top peak of the displayed signal is as near the screen top as possible, but still below it. Adjust the timebase switch so that about five cycles of the signal are displayed across the screen.

Next select the vertical cursor mode on your oscilloscope. First set the lower, or cursor 1 line to the bottom peaks of the displayed signal. Then set the upper, or cursor 2 line exactly on the top peaks of the sinewave, and note the amplitude displayed on the ΔV readout. Figure 16.1 shows the result of these settings. The amplitude readout was 1.32 V. Note this value of 1.32 and move the upper cursor down to the vertical centre of the sinewave signal so that the readout indicates half the previous value. Here the value is 0.66 V. Now leave the upper cursor set to this point and move the lower cursor down to align exactly with the bottom graticule line, or baseline. Then read the d.c. offset value as the ΔV value displayed on the readout. Figure 16.2 shows the signal in the example with the cursors set, and a readout value of 2.38 V, which is the d.c. offset.

If your signal has a negative d.c. offset, then initially set the baseline on the top graticule line with the coupling switch set to ground, and make your measurements relative to the top line.

Returning to our example, if the d.c. offset is large compared to the

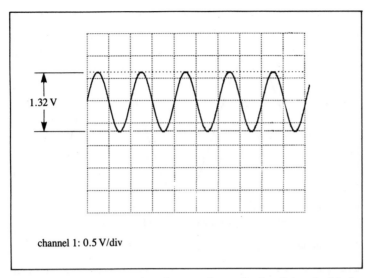

Figure 16.1. Cursor measurement of peak-to-peak amplitude of signal with d.c. offset.

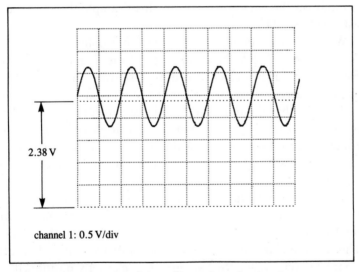

Figure 16.2. Cursor measurement of d.c. offset value of a signal.

sinewave amplitude, it will be necessary now to measure the peak-to-peak voltage more accurately. Set the Channel 1 input coupling switch to a.c., and use the Channel 1 position control to set the display at the centre of the screen. Turn the attenuator switch further clockwise to display the signal as large as possible without exceeding the screen area. Move the upper cursor to exactly touch the top tips of the sinewave signal, and align the lower cursor line to the bottom tips of the signal. You can then read the exact amplitude as the ΔV

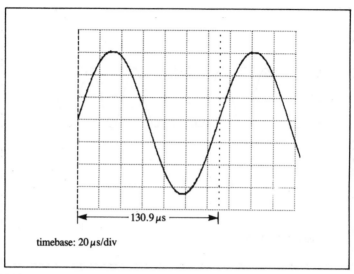

timebase: 20 μs/div

Figure 16.3. Cursor measurement of period time of a sinewave signal.

readout on the cursor display. In this example the signal amplitude was read as 1.34 V.

We now know that our signal has an amplitude of 1.34 V, and a d.c. offset of +2.38 V. Next, to measure the frequency, turn the timebase switch further clockwise until just over one complete cycle is displayed across the screen. Keep the input coupling switch set to a.c., and ensure that the timebase VARiable control is in the CALibrated position. Then using the Channel 1 position control and the horizontal position control, adjust the trace so that the waveform start is exactly half-way up the first vertical graticule line on the left (see Fig. 16.3).

Select the horizontal cursor mode and set the left cursor exactly on the first graticule line on the left where the trace starts. Move the right-hand cursor to the start of the second cycle of the waveform. This is where the trace crosses the graticule centre line after one complete cycle. When the two cursors are exactly positioned, read the Δt display to give the period of the waveform. In this case, the time was 130.9 μs. Select the frequency function on the readout display to give the signal frequency directly as 7.639 kHz. If you do not have this f function on your model, then calculate the frequency from $f = 1/t$ as usual.

We now have all the information possible about the input signal. As you make each type of measurement, always make each parameter as large as possible on the screen but within the screen area. This will give the greatest accuracy and fully utilize the cursor system.

INDEX